# Gravitational Lensing in
# COSMOLOGY

# Gravitational Lensing in
# COSMOLOGY

## Toshifumi Futamase

*Kyoto Sangyo University, Japan*

**World Scientific**

NEW JERSEY · LONDON · SINGAPORE · BEIJING · SHANGHAI · HONG KONG · TAIPEI · CHENNAI · TOKYO

*Published by*

World Scientific Publishing Co. Pte. Ltd.

5 Toh Tuck Link, Singapore 596224

*USA office:* 27 Warren Street, Suite 401-402, Hackensack, NJ 07601

*UK office:* 57 Shelton Street, Covent Garden, London WC2H 9HE

Library of Congress Control Number: 2024950369

**British Library Cataloguing-in-Publication Data**
A catalogue record for this book is available from the British Library.

**GRAVITATIONAL LENSING IN COSMOLOGY**

ISBN 978-981-327-678-9 (hardcover)
ISBN 978-981-327-679-6 (ebook for institutions)
ISBN 978-981-327-680-2 (ebook for individuals)

For any available supplementary material, please visit
https://www.worldscientific.com/worldscibooks/10.1142/11169#t=suppl

Typeset by Stallion Press
Email: enquiries@stallionpress.com

# Contents

# Chapter 1

# Brief Introduction to General Relativity

The theoretical basis of the observational cosmology is general relativity. The main topic in this book is gravitational lensing and thus the knowledge of relativity will help to understand the subject. General relativity reduces to Newtonian gravity in an appropriate limit called Newtonian limit where gravity is weak and the typical velocity of the system in question is much smaller than the speed of light $c$. The weakness of gravity is represented by the parameter $\frac{GM}{c^2R} \ll 1$, where $G$ is the gravitational constant, $M$ typical mass and $R$ typical length of the system in question. The gravitational constant is one of the most fundamental constants of nature that has dimensions.

$$G = 6.67430 \times 10^{-11} \mathrm{m}^3 \mathrm{kg}^{-1} \mathrm{s}^{-2} \tag{1.0.1}$$

General relativity is built on Newton gravity, so let us start with Newton gravity.

## 1.1. Newtonian Gravity

Newton derived the equation of motion based on the law of inertia which says that the state of motion does not change unless a force is applied

$$\boldsymbol{a} = \frac{\boldsymbol{F}}{m_I} \tag{1.1.1}$$

1

where $\boldsymbol{a}$ is the rate of change of velocity, namely, the acceleration. $m_I$ is the inertia mass. Therefore, the inertial mass is the indicator of the difficulty of changing motion.

Here, we are interested in gravity as a force. Newton derived the law of gravity between two point masses with masses $M_G$, and $m_G$ separated by a vector $\boldsymbol{r}$ as follows.

$$\boldsymbol{F}(\boldsymbol{x}) = -G\frac{M_G m_G}{r^2}\boldsymbol{n} \qquad (1.1.2)$$

where $r = |\boldsymbol{r}|$ is the distance between two point masses, and $\boldsymbol{n} = \frac{\boldsymbol{r}}{r}$ the unit vector along the direction $\boldsymbol{r}$. We define the positive direction as the direction increasing the distance. The masses $M_G, m_G$ are the gravitational masses which are the measure of exerting or receiving gravity. Since this equation is symmetric in both masses, we do not need to distinguish them.

Consider the case $M_G \gg m_G$. Then we may regard that the mass $m_G$ is subject to gravity generated by the mass $M_G$. Then Eq. (1.1.2) may be written as follows.

$$\boldsymbol{F}(\boldsymbol{x}) = -m_G \nabla \phi(\boldsymbol{x}) \qquad (1.1.3)$$

with

$$\phi(\boldsymbol{x}) = -\frac{GM_G}{r} \qquad (1.1.4)$$

where $\nabla$ is the derivative operator with respect to the directional vector in Cartesian coordinates. Equation (1.1.3) allows us to interpret the quantity $\phi$ as a potential created by the mass $M_G$. This potential is called gravitational or the Newtonian potential.

In this way, we obtain the equation of motion for a mass $m_G$ in the gravitational field created by a mass $M_G$ as

$$\boldsymbol{a} = -\frac{m_G}{m_I}\nabla \phi \qquad (1.1.5)$$

Our experience as well as all the experiments so far show that the falling motion does not depend on its mass (equivalence principle). This leads to the conclusion that the ration $m_G/m_I$ is a constant which we can choose the unity by appropriately choosing unit. Thus, the equivalence principle may be regarded as the equality between the inertia mass and gravitational mass. This is not an obvious fact since the physical concept between two types of mass is completely different. One can better to name the gravitational mass as gravitational charge in order to distinguish clearly two concepts. Henceforth, we do not distinguish two types of mass and call simply "mass".

The astronomical objects we consider here consist of huge number of particles and thus we need the expression for the gravitational force exhibited by such extended sources. Suppose the object is composed of $N$ particles with mass $M_A$ at $r_A$ where $A = 1, 2, \ldots, N$. Then the Newtonian potential by these particles are written as the sum of the potential of each particle.

$$\phi(r) = -\sum_{A=1}^{N} \frac{GM_A}{r_A} \tag{1.1.6}$$

where $r_A = r - r_A$ is the vector from the position of particle $A$ to the particle at position $r$ exerted by gravity, and $r_A = |r_A|$.

The extended object may be regarded that the mass is continuously distributed rather than consisted of many particles. In that case, the mass $M_A$ is regarded as the mass contained in the infinitesimally small volume element $dV = d^3r'$. Thus, the mass can be expressed by introducing the mass density at $r'$ as $M_A = \rho(r')d^3x'$. This allows us to write the Newtonian potential as follows.

$$\phi(x) = -G \int d^3x' \frac{\rho(x')}{|r - r'|} \tag{1.1.7}$$

The potential has a time dependence from the time dependence of the mass density. We have not indicated the time dependence.

Since the gravity is not the action at a distance force, it is better to express the equation for the law of gravity as a differential equation. In fact the above integral form needs to an appropriate boundary condition at spatial infinity. In deriving the equation, we remind the formula.

$$\Delta \frac{1}{|\boldsymbol{r}|} = -4\pi\delta(\boldsymbol{r}) \tag{1.1.8}$$

where, $\delta(\boldsymbol{r})$ is Dirac delta function in three-dimension, and $\Delta$ is the Laplacian defined by

$$\Delta = \nabla \cdot \nabla = \frac{\partial^2}{\partial x^2} + \frac{\partial^2}{\partial y^2} + \frac{\partial^2}{\partial z^2} \tag{1.1.9}$$

Sifting the origin we obtain

$$\Delta \frac{1}{|\boldsymbol{x} - \boldsymbol{x}'|} = -4\pi\delta(\boldsymbol{x} - \boldsymbol{x}') \tag{1.1.10}$$

Thus, we apply the Laplacian operator to the both side of Eq. (1.1.7)

$$\Delta\phi(\boldsymbol{x}) = -G \int d^3x' \rho(\boldsymbol{x}') \Delta \frac{1}{|\boldsymbol{x} - \boldsymbol{x}'|} \tag{1.1.11}$$

Finally, we obtain the differential equation satisfied by the Newtonian potential.

$$\Delta\phi(\boldsymbol{x}) = 4\pi G\rho(\boldsymbol{x}) \tag{1.1.12}$$

This equation is called Poisson equation and the basic equation in Newtonian gravity.

## 1.2.  Breakdown of Newtonian Gravity

Newtonian gravity explained not only the observed motion of planets in the solar system beautifully but also succeeded to predict the existance of Neptune and Pluto. A sign of its breakdown began to appear at the end of the 19th century as the observation of perihelion shift anomaly of Mercury. Mercury is the closest to the Sun whose

orbit is an ellipse with a large ellipticity[1] $e = 0.21$ (compare this with
Earth's value $e = 0.0167$). The position of the perihelion (the closest
position to the Sun) shift slightly in each cycle of the orbital motion.
This shift is measured by the change in angle seen from the focal
point, the Sun, and is 574.1 arc seconds in 100 years. Venus, which
is the closest to Mercury, makes the largest contribution, accounting
for 277.8 arc seconds, followed by the gravity of Jupiter, which has
the largest mass among planets in solar system, at 153.6 arc seconds,
and then Earth, which accounts for 90.0 arc seconds. Even if the
contributions of other planets are combined, the result is 531.2 arcsec,
and the remaining 42.9 arcsec cannot be explained in Newtonian
gravity. This is called Mercury's perihelion anomaly.

Not only that, the conceptual bases of Newtonian gravity was
broken. Namely, Newton gravity is based on the absolute space
and the absolute time which are denied by special relativity. The
observation that the speed of light does not change for any inertial
observer makes it clear that space and time for a given observer are
nothing more than the coordinate system of a certain 4-dimensional
space called spacetime. The system that assigns four real values
$X^\mu$, $\mu = 0, 1, 2, 3$ to a point in the spacetime (called an event) is
called a coordinate system. In particular $X^0$ is the time coordinate,
and sometimes written a $X^0 = cT$, where $c$ is the speed of light.
We use the unit $c = 1$ in almost cases. An inertial coordinate
system or simply inertial frame is a particular coordinate system
in which the law of inertia holds. The universality of the speed of
light is implemented in the spacetime by the requirement that the
following combination of the coordinate difference $dX^\mu$ between two
neighboring events does not depend on the inertial coordinate chosen.

$$ds^2 = -(dX^0)^2 + (dX^1)^2 + (dX^2)^2 + (dX^3)^2 \tag{1.2.2}$$

---

[1]The ellipticity of the orbit with the length of major and minor axis $2a$, $2b$,
respectively is defined as follows.

$$\epsilon = \sqrt{\frac{a^2 - b^2}{a^2}} \tag{1.2.1}$$

This combination is called the line element. Spacetime characterized by the above line element is called Minkowski spacetime.

We write the line element in the following form.

$$ds^2 = \eta_{\mu\nu}dX^\mu dX^\nu \tag{1.2.3}$$

Here, the repeated indices such as $\mu$ means that we sum over that indices from 0 to 3. This rule is called Einstein summation rule, and the indices summed over are called the dummy indices since the subscript name does not matter in the result. The quantity $\eta_{\mu\nu}$ is called Minkowski metric tensor. When $\eta_{\mu\nu}$ is regarded $4 \times 4$ metric, then the non-vanishing components of the matrix are only the diagonal components and sometimes written as

$$\eta = \text{diag}(-1, 1, 1, 1) \tag{1.2.4}$$

In the following, the coordinate transformation plays an important role. Two different inertial frames, say $O$ and $O'$ give different coordinates $X^\mu$ in $O$ and $X^{\mu'}$ in $O'$ for the same event. Since they express the same event, they are related with some linear relation.

$$X^{\mu'} = \Lambda^{\mu'}_{\ \nu}X^\nu \tag{1.2.5}$$

where the coefficients $\Lambda^{\mu'}_{\ \nu}$ are some constant because both coordinate are inertial frames.

Suppose we consider two neighboring events $P$ and $Q$. In the frame $O$ they have the coordinate $X^\mu$ and $X^\mu + dX^\mu$. The same events have coordinates $X^{\mu'}$ and $X^{\mu'} + dX^{\mu'}$ in the frame $O'$. The coordinate difference between two events are $dX^\mu$ and $dX^{\mu'}$ in the frame $O$ and $O'$, respectively. Then these coordinate differences satisfy the same relationship as the above equation (1.2.5), and thus the transformation matrix satisfies the following relation.

$$\eta_{\mu\nu}\Lambda^{\mu'}_{\ \alpha}\Lambda^{\nu'}_{\ \beta} = \eta_{\alpha\beta} \tag{1.2.6}$$

The transformation between any two inertial frames which satisfies the above relation is called Lorentz transformation. This is easily derived by writing the line element as follows.

$$ds^2 = \eta_{\mu\nu}dX^\mu dX^\nu = \eta_{\alpha\beta}dX^{\alpha'}dX^{\beta'} \tag{1.2.7}$$

## 1.3. Gravity as an Acceleration

The first step to construct a relativistic theory of gravity, we remind the law of equivalence again since it says the equivalence between gravity and acceleration. To see this explicitly, consider the 1-dimensional motion with a constant acceleration in Minkowski spacetime.

In Newtonian mechanic, the trajectory of a particle is expressed as position at each time, namely the position is a function of time, and its derivative gives the velocity. In relativity the time is also one of the coordinate in spacetime. Therefore, we need a real parameter to specify a position(event) in spacetime. The trajectory of a particle in spacetime is called a worldline. Along the worldline we can define the propertime $\tau$ as follows.

$$d\tau \equiv \sqrt{-ds^2} = dX^0\sqrt{1 - \mathbf{V}^2} \tag{1.3.1}$$

where $\mathbf{V}^2 = (V^1)^2 + (V^2)^2 + (V^3)^2$ is the square of the magnitude of 3-dimensional velocity. In the limit, where the 3-dimensional velocity is sufficiently smaller than the speed of light (non-relativistic limit), $d\tau = dX^0$, and using $\tau$ as a parameter, the spatial component of the quantity defined above becomes a 3-dimensional velocity. When viewed from a system moving at the same speed as the particle (such a coordinate system is called a comoving coordinate system), the 3-dimensional velocity is 0, so the proper time itself is the coordinate time in the comoving frame.

Then it is natural to define 4-velocity and 4-acceleration of a worldline as follows.

$$U^\mu = \frac{dX^\mu}{d\tau} \tag{1.3.2}$$

$$a^\mu = \frac{dU^\mu}{d\tau} \tag{1.3.3}$$

By definition the 4-velocity satisfies $U^2 \equiv \eta_{\mu\nu}U^\mu U^\nu = -1$. The following relation is useful to remember.

$$U^0 = \frac{1}{\sqrt{1 - V^2}}, \quad U^i = U^0 V^i \tag{1.3.4}$$

Four-dimensional momentum is defined as follows.

$$p^\mu = mU^\mu \tag{1.3.5}$$

where $m$ is the rest mass.

## Remark of notation

**1.** A 4-dimensional vector is defined as a 4-components quantity that has the same coordinate transformation properties as the coordinates. Thus, the 4-velocity and 4-acceleration are vector. A vector itself is coordinate free concept, but its components depend on the coordinate chosen. In order to make this clear we sometimes use the following notation.

$$\vec{A} \underset{O}{\rightarrow} (A^\mu) \tag{1.3.6}$$

This says that $(A^\mu)$ are components of a vector $\vec{A}$ in a coordinate $O$.

**2.** We also define the inner product of two vector $\vec{A}$ and $\vec{B}$ as follows.

$$\vec{A} \cdot \vec{B} = \eta_{\mu\nu}A^\mu B^\nu \tag{1.3.7}$$

In particular the inner product of the same vector is called as the magnitude of the vector. Note that the magnitude is not positive definite.

When $\vec{A} \cdot \vec{B} = 0$, then these vectors are said to be orthogonal. By the definition of 4-velocity, the 4-velocity and 4-acceleration of a worldline is orthogonal.

$$\vec{a} \cdot \vec{U} = 0 \tag{1.3.8}$$

A 1-component quantity which does not depend on the choice of coordinate is called scalar. The inner product of two vectors is a scalar.

**3.** A vector is called timelike when its magnitude is negative. Thus the 4-velocity is a timelike vector. A vector is called spacelike when its magnitude is positive. Thus, 4-acceleration is a spacelike vector. A vector is called null when its magnitude vanishes. The worldline of a photon is has a null tangent vector. Thus, the photon does not have its proper time. When the worldline of a photon is expressed as $X^\mu = X^\mu(\lambda)$ with some appropriate real parameter $\lambda$, then the 4-momentum of photon is defined

$$\vec{k} \underset{O}{\to} (k^\mu) = \left(\frac{dX^\mu}{d\lambda}\right) \tag{1.3.9}$$

with $\vec{k} \cdot \vec{k} = 0$.

**4.** We define a 4-component quantity with lower index by the following relation.

$$A_\mu = \eta_{\mu\nu} A^\nu \tag{1.3.10}$$

This quantity is called 1-form and it transforms oppositely as a vector. This can be understood by noting that the scalar product is a scalar.

$$\vec{A} \cdot \vec{B} = A_\mu B^\mu = A_{\alpha'} B^{\alpha'} = A_{\alpha'} \Lambda^{\alpha'}_\mu B^\mu \tag{1.3.11}$$

The operation of aligning the superscript and subscript and calculating the sum from 0 to 3 for those subscripts is called contraction. Note that in the case of contractions, any subscript represents the same content.

$$A^\mu B_\mu = A^\rho B_\rho \tag{1.3.12}$$

**5.** The concept of vector and 1-form may be generalized to tensor. A $(0, 2)$ tensor is defined as a 16-component quantity with two subscripts such that its contraction with a vector gives a 1 form. Similarly, a $(1, 1)$ tensor is a 16-component quantity with one subscript and one superscript such that its contraction with a vector gives a vector and its contraction with a 1 form gives a 1 form, and a $(2,0)$ tensor is a 16-component quantity with two superscripts such that its contraction with 1 form gives a vector. Higher rank tensors are defined inductively from lower rank tensor. As an example, we consider the energy-momentum tensor $T$ which expresses the energy and kinetic state of a continuum. Consider a small region $D$ within the continuum where the motion can be treated as a unit. Thus, the region has a 4-velocity $\vec{U}$. Then one can define the energy density measured in the rest frame of the region as

$$\rho = T_{\mu\nu}U^{\mu}U^{\nu} \tag{1.3.13}$$

We now consider 3-dimensional space which is orthogonal to the four velocity $\vec{U}$. This 3-dimensional space is the space where the region is at rest and we define that the only non-vanishing components of the energy-momentum tensor is isotropic pressure $P$. Namely, we set

$$T_{\mu\nu} = Ph_{\mu\nu} \tag{1.3.14}$$

where $h_{\mu\nu} = \eta_{\mu\nu} + U_{\mu}U_{\nu}$ that satisfies $h_{\nu\nu}U^{\mu} = 0$. In fact, $\vec{U}$ has components $(1,0,0,0)$ in the rest frame. So, in this frame $h_{\mu\nu}$ has only diagonal components $(0,1,1,1)$. Thus, $T_{ij} = P\delta_{ij}$ as expected. The fluid with this form of energy-momentum tensor is called a perfect fluid where there are no internal heat transfer and no friction, The energy-momentum tensor of a perfect fluid is written in general coordinate system as follows.

$$T^{\mu\nu} = \rho U^{\mu}U^{\nu} + Ph^{\mu\nu} \tag{1.3.15}$$

The law of energy conservation and momentum conservation are combined into one equation as follows.

$$T^{\mu\nu}_{,\nu} = 0 \tag{1.3.16}$$

### 1.3.1. *1-dimensional constant acceleration motion*

Let come back to a 1-dimensional acceleration motion with a constant acceleration

$$a^2 = -(a^0)^2 + (a^1)^2 = g^2 \qquad (1.3.17)$$

where $g$ is a constant. Thus, the components are expressed as follows.

$$a^0 = g \sinh(\tau), \quad a^1 = g \cosh(\tau) \qquad (1.3.18)$$

Then the 4-velocity is

$$U^0(\tau) = \cosh(g\tau), \quad U^1(\tau) = \sinh(g\tau) \qquad (1.3.19)$$

Integrating over the proper time gives

$$X^0(\tau) = \frac{1}{g}\sinh(g\tau), \quad X^1(\tau) = \frac{1}{g}\cosh(g\tau) \qquad (1.3.20)$$

where we have assumed $X^0 = 0$ at $\tau = 0$. Therefore, the worldline of this motion is the hyperbola.

$$-\left(X^0\right)^2 + \left(X^1\right)^2 = \frac{1}{g^2} \qquad (1.3.21)$$

This asymptotes to $X^0 = -X^1$ at $\tau \to -\infty$ on the $X^1 > 0$ plane, crosses the $X^1$ axis at $\frac{1}{g}$ at $X^0 = 0$, and asymptotically approaching $X^0 = X^1$ at $\tau \to \infty$ (on the $X^1 < 0$ plane, this hyperbola is reversed with respect to the $X^1 = 0$ axis. Here, we will limit the discussion to the $X^1 > 0$ plane.) The 3-dimensional speed of this motion is

$$V = \frac{U^1}{U^0} = \tanh(g\tau) \qquad (1.3.22)$$

Therefore, it approaches the origin from infinity at the speed of light while decelerating, and at $\tau = 0$ the speed becomes 0, then reverses and accelerates, approaching the speed of light at infinity.

To see gravity as an apparent force by this accelerated motion, we take time coordinate as the proper time $\tau$ and the spatial coordinate as $\frac{1}{g} + x$. Therefore, an observer moving with acceleration $g$ always remains at the spatial origin $x = 0$. To make a discussion clear we change the notation as $T = X^0$, $X = X^1$. The worldline $x = $ constant. is the trajectory of the observer with constant acceleration $\frac{g}{1+gx}$.

$$T(\tau, x) = \left(x + \frac{1}{g}\right)\sinh(g\tau), \quad X(\tau, x) = \left(x + \frac{1}{g}\right)\cosh(g\tau)$$

$$(1.3.23)$$

or, equivalently

$$\tau = \frac{1}{g}\tanh^{-1}\frac{T}{X}, \quad x = \sqrt{X^2 - T^2} - \frac{1}{g} \quad (1.3.24)$$

The line element (we have ignored other spatial direction except $x$-axis in the line element) may be written in both coordinates as follows.

$$ds^2 = -dT^2 + dX^2 = -(1 + gx)^2\,d\tau^2 + dx^2 \quad (1.3.25)$$

In general, we write the line element expressed in terms of general coordinates not necessarily inertial as follows.

$$ds^2 = g_{\mu\nu}(x)dx^\mu dx^\nu \quad (1.3.26)$$

where $g_{\mu\nu}$ is called the metric tensor in general coordinates.

Let write down uniform linear motion in this coordinate system $(\tau, x)$. In the original inertial frame $(T, X)$, the motion is simply

$$\frac{dU^\alpha}{d\tau} = 0 \quad (1.3.27)$$

We write a vector in the inertia frame using uppercase letters such $U^\alpha$, and a vector in the $(\tau, x)$ coordinate system using lower case letters such as $u^\mu$. Then the 4-dimensional velocity is converted as follows.

$$U^\alpha = \frac{\partial X^\alpha}{\partial x^\mu}u^\mu \quad (1.3.28)$$

Therefore, the uniform linear motion is written as

$$\frac{d}{d\tau}\left(\frac{\partial X^\alpha}{\partial x^\mu}u^\mu\right) = \frac{\partial X^\alpha}{\partial x^\mu}\frac{du^\mu}{d\tau} + u^\mu\frac{d}{d\tau}\left(\frac{\partial X^\alpha}{\partial x^\mu}\right) = 0 \quad (1.3.29)$$

where $\frac{\partial X^\alpha}{\partial x^\mu}$ is a function of $x$, and thus the $\tau$ derivative is derivative along the worldline.

$$\frac{d}{d\tau} = \frac{dx^\nu}{d\tau}\frac{\partial}{\partial x^\nu} = u^\nu\frac{\partial}{\partial x^\nu} \tag{1.3.30}$$

Finally, we obtain the equation of motion in the $(\tau, x)$ coordinate as follows.

$$\frac{du^\mu}{d\tau} + u^\nu u^\rho \frac{\partial^2 X^\alpha}{\partial x^\nu \partial x^\rho}\frac{\partial x^\mu}{\partial X^\alpha} = 0 \tag{1.3.31}$$

Now, we calculate the second term in the left-hand side of the above equation. For simplicity, we restrict ourself into the case where the velocity is much smaller than the speed of light. Thus, the 4-velocity is approximated as $\vec{u} = (1, v), |v| \ll 1)$. Then the spatial part of Eq. (1.3.31) may be written as

$$\frac{dv}{d\tau} + \frac{\partial^2 X^\alpha}{\partial \tau^2}\frac{\partial x}{\partial X^\alpha} = 0 \tag{1.3.32}$$

Using the coordinate transformation (1.3.24), we obtain, for example

$$\frac{\partial x}{\partial T} = -\frac{T}{\sqrt{X^2 - T^2}} = -\frac{1}{1 + gx}\sinh g\tau, \quad \frac{\partial T}{\partial \tau} = \cosh g\tau \tag{1.3.33}$$

Finally, within our approximation we obtain

$$\frac{dv}{d\tau} = -\frac{g}{1 - gx} \simeq -g \tag{1.3.34}$$

This is nothing but the equation of motion in Newtonian gravity. In this way, we recover the gravity from the accelerated motion.

## 1.4. Limitation of Equivalence Principle and Curved Spacetime

We have seen that Newtonian gravity recovered by an accelerated motion. However, this is not the whole story because the equivalence principle holds only in a sufficiently small region in spacetime. In fact the relative distance between two particles will change in time during the motion in a general gravitational field because of tidal effect. The equivalence principle holds true only in regions of spacetime where spatial and temporal changes in the gravitational field can be ignored. One can set up an inertial coordinate system within such a small region and such coordinate is called a locally inertial frame. It is impossible to cover the whole spacetime by one global inertial frame if there is a gravitational field. Einstein interpreted this to mean that spacetime is curved. Therefore, the curvature of spacetime is equivalent to gravitational field.

We will give a full description of the curvature of space-time later, but first we will explain the relationship between the inertial frame and the general coordinate system. To make the discussion clear we write the coordinates in a local inertial frame $O$ as $(X^{\bar{\alpha}})$ where we used uppercase letter as well as added a bar to the subscripts. On the other hand, we write the coordinates in a general frame $S$ as $(x^{\mu})$ as usual. The relation between two frames is expressed the following quantity.

$$e^{\bar{\alpha}}_{\mu}(x) \equiv \frac{\partial X^{\bar{\alpha}}}{\partial x^{\mu}} \qquad (1.4.1)$$

where $e^{\bar{\alpha}}_{\mu}$ are the function of $x^{\mu}$. A connection between nearby two inertial frame may be described by the difference of this quantity between two neighboring events, and we write the difference as

$$e^{\bar{\alpha}}_{\mu}(x + dx) - e^{\bar{\alpha}}_{\mu}(x) = \Gamma^{\rho}_{\mu\nu}(x)e^{\bar{\alpha}}_{\rho}(x)dx^{\nu} \qquad (1.4.2)$$

This is the definition of the quantity $\Gamma^{\rho}_{\mu\nu}(x)$. Namely,

$$\frac{\partial e^{\bar{\alpha}}_{\mu}}{\partial x^{\nu}} = \Gamma^{\rho}_{\mu\nu}e^{\bar{\alpha}}_{\rho} \qquad (1.4.3)$$

This function $\Gamma$ is called Christoffel symbol. One can see that Christoffel symbol is symmetric with respect to the exchange of the two lower subscripts. Therefore, its independent components are 40. The Christoffel symbol is equivalently expressed in terms of the metric tensor $g_{\mu\nu}$ in general accelerated frame. Note that metric tensor in general frame is related with Minkowski metric as follows.

$$g_{\mu\nu}(x) = \eta_{\alpha\beta}e^{\bar{\alpha}}_{\mu}(x)e^{\bar{\beta}}_{\nu}(x) \tag{1.4.4}$$

This relation is obtained by noticing the invariance of the line element.

$$ds^2 = \eta_{\alpha\beta}dX^{\bar{\alpha}}dX^{\bar{\beta}} = g_{\mu\nu}(x)dx^{\mu}dx^{\nu} \tag{1.4.5}$$

By differentiating both eq. (1.4.4) we have

$$\frac{\partial g_{\mu\nu}}{\partial x^{\rho}} = \Gamma^{\sigma}_{\mu\rho}g_{\sigma\nu} + \Gamma^{\sigma}_{\nu\rho}g_{\mu\sigma} \tag{1.4.6}$$

From this we have the desired relation.

$$\Gamma^{\rho}_{\mu\nu} = \frac{1}{2}g^{\rho\sigma}\left(\frac{\partial g_{\sigma\nu}}{\partial x^{\mu}} + \frac{\partial g_{\mu\sigma}}{\partial x^{\nu}} - \frac{\partial g_{\mu\nu}}{\partial x^{\sigma}}\right) \tag{1.4.7}$$

### 1.4.1. *Covariant derivative and Geodesic equation*

Christoffel symbol gives us a well defined derivative operator in curved spacetime. Differentiation is basically taking the difference between two nearby points. In the case of Minkowski spacetime, this is simply taking a partial derivative because any quantity such as a vector transforms in the same way at any point and thus the difference is also a vector. On the other hand, the transformation property is different at each event in curved spacetime and thus the difference does not have any well defined transformation property. In order to define an differential operation on a vector, we recall that the partial derivative for a vector in a locally inertial frame does have well defined meaning, and thus we define the derivative operator $\nabla$ for a vector $\vec{V}$ as follows

$$\nabla\vec{V} \underset{O}{\rightarrow} \left(\frac{\partial V^{\bar{\alpha}}}{\partial X^{\bar{\beta}}}\right) \tag{1.4.8}$$

This components may be also expressed in general frame as

$$\frac{\partial V^{\bar{\alpha}}}{\partial X^{\bar{\beta}}} = e^{\nu}_{\bar{\beta}}\frac{\partial}{\partial x^{\nu}}\left(e^{\bar{\alpha}}_{\mu}V^{\mu}\right) = e^{\nu}_{\bar{\beta}}e^{\bar{\alpha}}_{\nu}\left(\frac{\partial V^{\nu}}{\partial x^{\nu}} + \Gamma^{\mu}_{\rho\nu}V^{\rho}\right) \tag{1.4.9}$$

where $e^\nu_{\bar\beta} = \frac{\partial x^\nu}{\partial X^{\bar\beta}}$. This motivates us to define the derivative of a vector in general coordinates as

$$\nabla_\nu V^\mu = \frac{\partial V^\mu}{\partial x^\nu} + \Gamma^\mu_{\rho\nu} V^\rho \tag{1.4.10}$$

This is called a covariant derivative of a vector.

Similarly, we can generalize the concept of the covariant derivative to a tensor. A $(n, m)$ rank tensor $\boldsymbol{T}$ is defined as an $4^{n+m}$-component quantity in an inertial coordinate that satisfy the following transformation property.

$$T^{\bar\alpha'_1 \cdots \bar\alpha'_n}_{\bar\beta'_1 \cdots \bar\beta'_m} = \Lambda^{\bar\alpha'_1}_{\bar\gamma_1} \cdots \Lambda^{\bar\alpha'_n}_{\bar\gamma_n} \Lambda^{\bar\delta_1}_{\bar\beta'_1} \cdots \Lambda^{\bar\delta_m}_{\bar\beta'_m} T^{\bar\gamma_1 \cdots \bar\gamma_n}_{\bar\delta_1 \cdots \bar\delta_m} \tag{1.4.11}$$

where $T^{\bar\gamma_1 \cdots \bar\gamma_n}_{\bar\delta_1 \cdots \bar\delta_m}$ and $T^{\bar\alpha'_1 \cdots \bar\alpha'_n}_{\bar\beta'_1 \cdots \bar\beta'_m}$ are the components in an inertial frame $O$ and $O'$, respectively. $\Lambda^{\bar\alpha'}_{\bar\beta}$ and $\Lambda^{\bar\delta}_{\bar\beta'}$ are Local Lorentz transformation from $O$ to $O'$ and its inverse, respectively. Thus, a $(1, 0)$ rank tensor is a vector and $(0, 1)$ rank tensor is a 1-form. From the definition, it is easy to derive the covariant derivative for a $(n, m)$ rank tensor in a general frame by noticing

$$T^{\bar\alpha_1 \cdots \bar\alpha_n}_{\bar\beta_1 \cdots \bar\beta_m} = e^{\bar\alpha_1}_{\mu_1} \cdots e^{\bar\alpha_n}_{\mu_m} e^{\nu_1}_{\bar\beta_1} \cdots e^{\nu_m}_{\bar\beta_m} T^{\mu_1 \cdots \mu_n}_{\nu_1 \cdots \nu_m} \tag{1.4.12}$$

For example, a covariant derivative of a $(2, 1)$ rank tensor $\boldsymbol{M}$ may be calculated as follows.

$$\nabla_\rho M^{\mu\nu}_\sigma = \partial_\rho M^{\mu\nu}_\sigma - \Gamma^\xi_{\rho\sigma} M^{\mu\nu}_\xi + \Gamma^\mu_{\rho\xi} M^{\xi\nu}_\sigma + \Gamma^\nu_{\rho\xi} M^{\mu\xi}_\rho \tag{1.4.13}$$

Note that the covariant derivative of metric tensor identically vanishes.

$$\nabla_\rho g_{\mu\nu} = 0 \tag{1.4.14}$$

Therefore, the covariant derivative of its inverse $g^{\mu\nu}$ also vanishes.

The equation of motion of a particle in a gravitational field described by the metric tensor $g_{\mu\nu}$ is derived in the same way as in the accelerated frame in Minkowski spacetime. The difference is that we here consider a locally inertial frame not globally defined inertial frame.

$$U^{\bar{\alpha}} = \frac{dX^{\bar{\alpha}}}{d\lambda} = \frac{\partial X^{\bar{\alpha}}}{\partial x^{\mu}}\frac{dx^{\mu}}{d\lambda} = e^{\bar{\alpha}}_{\mu}(x)u^{\mu} \qquad (1.4.15)$$

Namely, we have the following equation called geodesic equation.

$$\frac{du^{\mu}}{d\lambda} + \Gamma^{\mu}_{\rho\sigma}u^{\rho}u^{\sigma} = 0 \qquad (1.4.16)$$

The worldline $x^{\mu}(\lambda)$ whose 4-velocity satisfies the geodesic equation is called geodesic curve, or simply geodesic.

Geodesic equation (1.4.16) is also written as

$$u^{\nu}\nabla_{\nu}u^{\mu} = 0 \qquad (1.4.17)$$

The geodesic equation in a general coordinate system can be automatically found by converting the partial differential in the equation of motion in a local inertial frame into a covariant differential. In this way, if the physical laws in the local inertial frame, that is, in special relativity are known, the physical laws in a general coordinate system, that is, in curved spacetime can be found by replacing partial differentials with covariant differentials. For example, the conservation law of energy momentum tensor (1.3.16) can be written as follows in a curved spacetime.[2]

$$\nabla_{\nu}T^{\mu\nu} = 0 \qquad (1.4.19)$$

Also the symbol; is often used to represent covariant differentiation. Thus, the above conservation law can be also written as

$$T^{\mu\nu}{}_{;\nu} = 0 \qquad (1.4.20)$$

---

[2]However, there remains freedom to add quantities that become 0 in Minkowski spacetime, such as the curvature tensor described below. For example, as a geodesic equation, we could adopt the following equation.

$$u^{\nu}\nabla_{\nu}u^{\mu} + \alpha R^{\mu}{}_{\nu}u^{\nu} = 0 \qquad (1.4.18)$$

where $\alpha$ is an arbitrary constant. The case where there is no such direct interaction with curvature is called minimal coupling. It is generally difficult to determine $\alpha$ through experiments, and minimal coupling is usually assumed.

## Null geodesic equation

A photon is a massless particle and propagate with the speed of light. Therefore, $ds^2 = 0$ between any two events on the worldline. This means that there is no proper time. Instead of the proper time, we take a real number $\lambda$ as a parameter of the worldline of a photon and define the 4 momentum as follows.

$$\vec{k} = \frac{d\vec{X}}{d\lambda} \underset{O}{\rightarrow} \left(\frac{dX^\mu}{d\lambda}\right) \qquad (1.4.21)$$

Then

$$k^2 = \eta_{\mu\nu}k^\mu k^\nu = 0 \qquad (1.4.22)$$

thus 4 momentum of a photon is a null vector.

The energy of a photon with 4 momentum $\vec{k}$ measured by an observer with 4 velocity $\vec{U}$ is defined as follows.

$$E = -\vec{U} \cdot \vec{k} \qquad (1.4.23)$$

Since this is a scalar, its value is the same in any frame. You will agree with this definition when you evaluate it in the rest frame of the observer. In general inertial frame $O'$, the null condition (1.4.22) allows us to write down the 4 momentum of a photon as follows.

$$\vec{k} \underset{O'}{\rightarrow} (E, E\boldsymbol{n}) \qquad (1.4.24)$$

where $\boldsymbol{n}$ is a unit spatial vector that represents the direction of photon travel in 3-dimensional space.

## Gravitational redshift

It is useful to derive the gravitational redshift here. Our definition (1.4.23) gives the following equation as the ratio of the frequency of photon when emitted $\omega_{em}$ and the frequency received $\omega_{obs}$

$$\frac{\omega_{obs}}{\omega_{em}} = \frac{\vec{k} \cdot \vec{u}_{obs}}{\vec{k} \cdot \vec{u}_{em}} \qquad (1.4.25)$$

For example, consider an observer $\vec{u}_{\text{obs}} = (1,0,0,0)$ that is stationary in an inertial frame $O$ and a light source that is stationary in a general frame $S$ in a gravitational field given by $ds^2 = g_{\mu\nu}dx^\mu dx^\nu$. Then the normalization condition $u_{\text{em}}^2 = -1$ gives $\vec{u}_{\text{em}} = (1/\sqrt{-g_{00}(P_{\text{em}})},0,0,0)$. Thus, we find the following formula.

$$\frac{\omega_{\text{obs}}}{\omega_{\text{em}}} = \sqrt{-g_{00}(P_{\text{em}})} \tag{1.4.26}$$

This is known as gravitational redshift.

## 1.5. Curvature Tensor

Up until now, we have been using the term curved spacetime, but we have limited our discussion to the vicinity of a single event. Whether spacetime is truly curved cannot be determined only by the metric tensor nor Christoffel symbol. This is because locally we can always choose the flat metric where the Christoffel symbol vanishes. Note that Christoffel symbol is not tensor so that the disappearance of all components of the Christoffel symbol does not mean that spacetime is flat. We need a tensorial quantity to express that the spacetime is curved. To find such a quantity we go back to the situation where Einstein realized the equivalence between gravity and curvature and consider the difference in motion between the two events. The consideration in this situation leads to the tensor we are looking for.

Consider two nearby geodesics of two particles. Let the world line in the local inertial frame of one particle be $X^{\bar{\alpha}}(\lambda)$, and the other particle be an infinitesimal amount $\Xi^{\bar{\alpha}}$ away from that world line. Thus the worldline of the other particle may be described as, $\tilde{X}^{\bar{\alpha}} = X^{\bar{\alpha}} + \Xi^{\bar{\alpha}}$. Let us also take the corresponding worldlines in a general coordinate as $x^\mu(\lambda)$, $\tilde{x}^\mu(\lambda) = x^\mu(\lambda) + \xi^\mu(\lambda)$, respectively. Then the tangent vectors of two worldlines may be written as follows.

$$\vec{U}(x) = u^\mu(x)\vec{e}_\mu(x) = \frac{dx^\mu}{d\lambda}\vec{e}_\mu(x)$$
$$\vec{\tilde{U}}(\tilde{x}) = \tilde{u}^\mu(\tilde{x})\vec{e}_\mu(\tilde{x}) = \frac{d\tilde{x}^\mu}{d\lambda}\vec{e}_\mu(\tilde{x}) \tag{1.5.1}$$

Here we used the notation $\vec{e}_\mu = (e^{\tilde{a}}_\mu)$. Since each worldlines are geodesics, they satisy

$$\frac{d\vec{U}}{d\lambda} = 0, \quad \frac{d\vec{\tilde{U}}}{d\lambda} = 0 \tag{1.5.2}$$

Now calculate the difference of these two geodesis. Geodesic equatiion for the vector $\vec{\tilde{U}}(\tilde{x})$ is as follows.

$$0 = \frac{d\vec{\tilde{U}}(\tilde{x})}{d\lambda} = \frac{d}{d\lambda}\left[\frac{d\tilde{x}^\mu}{d\lambda}\vec{e}_\mu(\tilde{x})\right] = \frac{d^2\tilde{x}^\mu}{d\lambda^2}\vec{e}_\mu(\tilde{x}) + \frac{d\tilde{x}^\mu}{d\lambda}\frac{d\tilde{x}^\mu}{d\lambda}\vec{e}_{\mu,\nu}(\tilde{x})$$
$$\tag{1.5.3}$$

If we expand this equation to first order in the infinitesimal quantity $\xi$ around the world line $x^\mu(\lambda)$, we obtain the following equation using $\frac{d\vec{U}}{d\lambda} = 0$.

$$0 = \frac{d^2\xi^\mu}{d\lambda^2}\vec{e}_\mu + 2u^\mu\frac{d\xi^\nu}{d\lambda}\vec{e}_{\mu,\nu} + \frac{d^2x^\mu}{d\lambda^2}\xi^\nu\vec{e}_{\mu,\nu} + u^\mu u^\nu\xi^\rho\vec{e}_{\mu,\nu\rho} \tag{1.5.4}$$

In order to transform the equaton to the equation for $\vec{\Xi}$, we calculate the secnd order derivative of $\vec{\Xi}$.

$$\frac{d^2\vec{\Xi}}{d\lambda^2} = \frac{d^2}{d\lambda^2}\left(\xi^\mu\vec{e}_\mu\right)$$

$$= \frac{d^2\xi^\mu}{d\lambda^2}\vec{e}_\mu + 2u^\mu\frac{d\xi^\nu}{d\lambda}\vec{e}_{\mu,\nu} + \frac{d^2x^\mu}{d\lambda^2}\xi^\nu\vec{e}_{\mu,\nu} + \vec{e}_{\mu,\rho\nu}u^\mu u^\nu\xi^\rho \tag{1.5.5}$$

Using this equation and eq. (1.5.4), we obtain

$$\frac{d^2\vec{\Xi}}{d\lambda^2} = \left(\vec{e}_{\mu,\nu\rho} - \vec{e}_{\mu,\rho\nu}\right)u^\mu u^\nu\xi^\rho. \tag{1.5.6}$$

Since this is a vector equation, we can rewrite thus as follows.

$$\frac{d^2\vec{\Xi}}{d\lambda^2} = u^\mu u^\sigma\xi^\rho R^\nu_{\mu\sigma\rho}\vec{e}_\nu \tag{1.5.7}$$

Using the relation

$$\frac{d\vec{\Xi}}{d\lambda} = \left(u^\alpha\nabla_\alpha\xi^\nu\right)\vec{e}_\nu, \tag{1.5.8}$$

we finely obtain the equation for the difference vector of two nearby geodesics as follows.

$$u^\rho\nabla_\rho\left(u^\sigma\nabla_\sigma\xi^\mu\right) = R^\mu_{\nu\rho\sigma}u^\nu u^\sigma\xi^\rho \tag{1.5.9}$$

This us called the geodesic deviation equation.

The four rank tensor in the right hand side of the geodesic deviatiin equation is called as Riemann tensor and is defined as follows

$$[\partial_\sigma, \partial_\rho]\vec{e}_\mu \equiv \vec{e}_{\mu,\rho\sigma} - \vec{e}_{\mu,\sigma\rho} = R^\nu{}_{\mu\sigma\rho}\vec{e}_\nu \qquad (1.5.10)$$

Using the relation $\vec{A} = A^\mu \vec{e}_\mu$, the following equation is obtained for an arvitrary vector $\vec{A}$.

$$[\nabla_\sigma, \nabla_\rho]A^\mu = R^\mu{}_{\nu\rho\sigma}A^\nu \qquad (1.5.11)$$

The Riemann tensor is also called the curvature tensor and describes how spacetime is curved. The two geodesics remain parallel and spacetime is flat if and only if all components of the Riemann tensor are zero. As shown later only half of the components of the Riemann tensor vanish, so space-time can bend without matter.

The concrete form of the Riemann tensor can be easily calculated from the above definition and the definition of the Christoffel symbol $\partial_\nu e^{\bar{\alpha}}_\mu = \Gamma^\rho{}_{\mu\nu}e^{\bar{\alpha}}_\rho$ as follows.

$$R^\mu{}_{\nu\rho\sigma} = \partial_\rho\Gamma^\mu{}_{\sigma\nu} - \partial_\sigma\Gamma^\mu{}_{\nu\rho} + \Gamma^\mu{}_{\sigma\gamma}\Gamma^\gamma{}_{\rho\nu} - \Gamma^\mu{}_{\rho\gamma}\Gamma^\gamma{}_{\sigma\nu} \qquad (1.5.12)$$

## Symmetries of Riemann tensor

Since the Riemann tensor is a fourth-rank tensor, the number of independent components is $4^4 = 256$ if there is no symmetry between the subscripts. However, as described below, there are various symmetries, so the number of independent components becomes significantly reduces. These symmetries can be easily derived using (1.5.10). Eq. (1.5.10) can be written as follows.

$$R_{\mu\nu\rho\sigma} = \vec{e}_\mu \cdot (\vec{e}_{\nu,\sigma\rho} - \vec{e}_{\nu,\rho\sigma}) \qquad (1.5.13)$$

Here, the dot represents the inner product in a local inertial frame such that $\vec{e}_\mu \cdot \vec{e}_\nu = g_{\mu\nu}$. This immediately shows that $R_{\mu\nu\rho\sigma}$ is antisymmetric with respect to the exchange of subscripts $\rho, \sigma$.

$$R_{\mu\nu\rho\sigma} = -R_{\mu\nu\sigma\rho} \qquad (1.5.14)$$

When we fix the first subscript of (1.5.13) and cycle through the remaining three subscripts and take the sum,

$$R_{\mu\nu\rho\sigma} + R_{\mu\rho\sigma\nu} + R_{\mu\sigma\nu\rho} = \vec{e}_\mu \cdot (\vec{e}_{\nu,\sigma\rho} - \vec{e}_{\nu,\rho\sigma} + \vec{e}_{\rho,\nu\sigma} - \vec{e}_{\rho,\sigma\nu} + \vec{e}_{\sigma,\rho\nu} - \vec{e}_{\sigma,\nu\rho})$$

Here, from the symmetry about the subscripts under the Christoffel symbol,

$$\vec{e}_{\nu,\rho} = \vec{e}_{\rho,\nu} \qquad (1.5.15)$$

Using this, we obtain the following identity

$$R_{\mu\nu\rho\sigma} + R_{\mu\rho\sigma\nu} + R_{\mu\sigma\nu\rho} = 0 \qquad (1.5.16)$$

This formula can also be written as:

$$R_{\mu[\nu\rho\sigma]} = 0 \qquad (1.5.17)$$

Here, the symbol [] represents complete antisymmetry of the subscripts between them.

It is also antisymmetry for exchanging the first two of the four subscripts. In fact,

$$R_{\mu\nu\rho\sigma} + R_{\nu\mu\rho\sigma} = \vec{e}_\mu \cdot \vec{e}_{\nu,\sigma\rho} + \vec{e}_\nu \cdot \vec{e}_{\mu,\sigma\rho} - (\vec{e}_\mu \cdot \vec{e}_{\nu,\rho\sigma} + \vec{e}_\nu \cdot \vec{e}_{\mu,\rho\sigma})$$
$$= (\vec{e}_\mu \cdot \vec{e}_\nu)_{,\sigma\rho} - (\vec{e}_\mu \cdot \vec{e}_\nu)_{,\sigma\rho} = 0$$

Thus, we have

$$R_{\mu\nu\rho\sigma} = -R_{\nu\mu\rho\sigma} \qquad (1.5.18)$$

It is easy to show the following symmetry from the above symmetries.

$$R_{\mu\nu\rho\sigma} = R_{\rho\sigma\mu\nu} \qquad (1.5.19)$$

## Ricci tensor and Ricci scalar

A second-rank tensor that is obtained by contracting the first and third subscripts of a Riemann tensor is called a Ricci tensor, and a scalar that is further contracted is called a Ricci scalar.

$$R_{\mu\nu} = R^{\rho}{}_{\mu\rho\sigma} = g^{\rho\sigma} R_{\rho\mu\sigma\nu} \tag{1.5.20}$$

$$R = R^{\mu}{}_{\mu} = g^{\mu\nu} R_{\mu\nu} \tag{1.5.21}$$

Ricci tensor is a symmetric tensor.

$$R_{\mu\nu} = R_{\nu\mu} \tag{1.5.22}$$

## Number of independent components of Riemann tensor

Let us now calculate the number of independent components of the Riemann tensor from the above symmetries. Here, the dimension of spacetime is assumed to be $N$ not necessarily 4.

We write the pair of first two subscripts as A and the pair of last two subscripts as B of the Riemann tensor such that $R_{AB}$ with two subscripts A and B. The independent number of each subscripts is $N(N-1)/2$ because each of them consists of an antisymmetric pair. Therefore, the number of the components of $R_{AB}$ is $(N(N-1)/2)^2$, but it becomes even less because of the symmetry $R_{\mu[\nu\rho\sigma]} = 0$. The number of this symmetry is given by $_NC_3$ because there are $N$ ways to take the first subscript $\mu$, and the remaining three subscripts cannot be the same. Therefore, the number of independent components of the Riemann tensor is

$$\left(\frac{N(N-1)}{2}\right)^2 - N\frac{N(N-1)(N-2)}{3!} = \frac{1}{12}N^2(N^2-1) \tag{1.5.23}$$

In 4-dimensional spacetime the independent number is 20. When these 20 components are all 0, the spacetime is a flat Minkowski spacetime.

The independent number of Ricci tensor is given by

$$N^2 - \frac{N(N-1)}{2} = \frac{N(N+1)}{2} \tag{1.5.24}$$

Ricci tensor has 10 independent components in 4-dimensional space-time.

## Weyl tensor

In the $N = 2$ dimension, the number of independent components of the Riemann tensor is 1, so the Riemann tensor tensor is expressed in terms of Ricci scalar $R$, and by considering the symmetry of the subscripts of the Riemann tensor, it takes the following form.

$$R_{ijk\ell} = \frac{1}{2}(g_{ik}g_{j\ell} - g_{i\ell}g_{jk})R \qquad (1.5.25)$$

In the $N = 3$ dimension, the number of independent components of the Riemann tensor and Ricci tensor are the same, therefore the Riemann tensor may be expressed in terms of the Ricci tensor and Ricci scalar as follows.

$$R_{ijk\ell} = g_{ik}R_{j\ell} - g_{i\ell}R_{jk} - g_{jk}R_{i\ell} + g_{j\ell}R_{ik} - \frac{1}{2}(g_{ik}g_{j\ell} - g_{i\ell}g_{jk})R$$
$$(1.5.26)$$

In $N \geq 4$ dimensions, the number of independent components of the Riemann tensor is larger than the number of independent components of the Ricci tensor. Therefore, the Riemann tensor has components that are not represented by the Ricci tensor. The component is called the Weyl tensor and is expressed as follows.

$$C_{\alpha\beta\mu\nu} = R_{\alpha\beta\mu\nu} - \frac{1}{N-2}\left(g_{\alpha\mu}R_{\beta\nu} - g_{\alpha\nu}R_{\beta\mu} - g_{\beta\mu}R_{\alpha\nu} + g_{\beta\nu}R_{\alpha\mu}\right)$$
$$+ \frac{R}{(N-1)(N-2)}\left(g_{\alpha\mu} - g_{\alpha\nu}g_{\beta\mu}\right) \qquad (1.5.27)$$

The Weyl tensor satisfies the following conditions by its construction.

$$C^{\alpha}{}_{\beta\alpha\nu} = 0 \qquad (1.5.28)$$

The number of independent components of the Weyl tensor can be determined by subtracting the number of independent components of the Ricci tensor from the number of independent components of the Riemann tensor.

$$\frac{1}{12}N^2(N^2-1) - \frac{1}{2}N(N+1) = \frac{1}{12}N(N+1)(N+2)(N-3)$$

$$(1.5.29)$$

A necessary and sufficient condition for spacetime to be flat is that all components of the Riemann tensor are 0. On the other hand, as we will see later, if there is no matter ($R_{\mu\nu} = 0$). This means that the Weyl tensor is not necessarily 0 even in the absence of matter. Therefore, Einstein's equations have non-trivial (meaning not flat) vacuum solutions. In fact, gravitational wave and black hole are the solutions of vacuum Einstein equation.

## Differential identity

The Riemann tensor has another important identity that involves differentiation. Define the commutator product for operators $A$, $B$ as follows

$$[A, B] \equiv AB - BA \qquad (1.5.30)$$

Then the following relation holds true as an identity.

$$[A, [B, C]] + [B, [C, A]] + [C, [A, B]] = 0 \qquad (1.5.31)$$

The identity is called Jacobi identity. Applying Jacobi identity to partial derivatives and then operates to $\vec{e}_\mu$, then we have

$$([\partial_\rho, [\partial_\mu, \partial_\nu]] + [\partial_\mu, [\partial_\nu, \partial_\rho]] + [\partial_\nu, [\partial_\rho, \partial_\mu]]) \, \vec{e}_\sigma = 0 \qquad (1.5.32)$$

For example, the first term of the left-hand side of this equation is calculated to be

$$[\partial_\rho, [\partial_\mu, \partial_\nu]]\vec{e}_\sigma = \partial_\rho(R^\lambda_{\ \sigma\mu\nu}\vec{e}_\lambda) - [\partial_\mu, \partial_\nu]\partial_\rho\vec{e}_\sigma$$

$$= \left(\partial_\rho R^\lambda_{\ \sigma\mu\nu} + R^\chi_{\ \sigma\mu\nu}\Gamma^\lambda_{\ \chi\rho} - \Gamma^\chi_{\ \rho\sigma}R^\lambda_{\ \chi\mu\nu}\right)\vec{e}_\lambda$$

Change partial derivatives to covariant derivatives gives

$$[\partial_\rho, [\partial_\mu, \partial_\nu]]\vec{e}_\sigma = \left(\nabla_\gamma R^\lambda{}_{\sigma\mu\nu} + \Gamma^\chi{}_{\mu\rho}R^\lambda{}_{\sigma\chi\nu} + \Gamma^\chi{}_{\nu\rho}R^\lambda{}_{\sigma\mu\chi}\right)\vec{e}_\lambda \quad (1.5.33)$$

Doing molar calculations and sum uo together, we find the following differential identity.

$$\nabla_\rho R^\lambda{}_{\sigma\mu\nu} + \nabla_\mu R^\lambda{}_{\sigma\nu\rho} + \nabla_\nu R^\lambda{}_{\sigma\rho\mu} = 0 \quad (1.5.34)$$

This identity is called Bianchi identity. Contracting the subscript $\lambda$ and $\mu$ in this identity, we get

$$\nabla_\rho R_{\sigma\nu} + \nabla_\mu R^\mu{}_{\sigma\nu\rho} - \nabla_\nu R_{\sigma\rho} = 0 \quad (1.5.35)$$

Furthermore, multiplying by $g^{\rho\sigma}$ and contracting gives the following identity, called the contracted bianchi identity.

$$\nabla_\rho \left(R^\rho_\nu - \frac{1}{2}\delta^\rho_\nu R\right) = 0 \quad (1.5.36)$$

The combination inside the parentheses is called the Einstein tensor.

$$G_{\mu\nu} \equiv R_{\mu\nu} - \frac{1}{2}g_{\mu\nu}R \quad (1.5.37)$$

## 1.6. Propagation of a Light Ray Bundle in a Curved Spacetime

As we saw above, in dimensions higher than 4, the Riemann tensor can be divided into the Ricci tensor and the Weyl tensor. The role of the Ricci tensor and Weyl tensor in curved spacetime can be understood by looking at how the cross-sectional of light ray changes during the propagation.

Consider light with a tiny cross section that propagates through spacetime. Since this light consists of countless rays, we call it a ray bundle. Apply the geodesic deviation equation to the two nearby rays that make up this ray bundle. Assuming that the 4-dimensional momentum of one ray is $\vec{k}$ and the vector connecting two rays with the same parameter value is $\vec{\xi}$, the geodesic deviation equation is written as follows:

$$\frac{D^2 \xi^\mu}{d\lambda^2} = R^\mu{}_{\nu\rho\sigma} k^\nu k^\rho \xi^\sigma \tag{1.6.1}$$

where $D/d\lambda$ corresponds to the derivative $d/d\lambda$ in a locl Lorentz frame is defined as

$$\frac{D}{d\lambda} = k^\mu \nabla_\mu \tag{1.6.2}$$

The four momentum of a photon satisfies the null geodesic equation.

$$k^\nu \nabla_\nu k^\mu = 0, \quad k^2 = g_{\mu\nu} k^\mu k^\nu = 0 \tag{1.6.3}$$

Now, we choose an orthonormal base vector $\vec{e}_{(a)}$, $a = 1,2$ in a 2-dimensional plane perpendicular to the direction of the photon path so that it moves in parallel along the photon.

$$\vec{k} \cdot \vec{e}_{(a)} = 0, \quad \vec{e}_{(a)} \cdot \vec{e}_{(b)} = \delta_{ab} \tag{1.6.4}$$

$$\frac{D}{d\lambda} \vec{e}_{(a)} = 0 \tag{1.6.5}$$

We also choose $\vec{\xi}$ in such a way that it is always perpendicular to $\vec{k}$. Thus, we can expand $\vec{\xi}$ by $\vec{e}_{(a)}$.

$$\xi^\mu(\lambda) = \sum_{a=1}^{2} \ell_a(\lambda) e^\mu_{(a)} \tag{1.6.6}$$

The expansion coefficients $\ell_a$ are scalars under the general coordinate transformation. We substitute the this expression into the geodesic deviation equation to get the following equation for the coefficients $\ell_a$.

$$\frac{d^2 \ell_a}{d\lambda^2} = \sum_b K_{ab} \ell_a \ell_b \tag{1.6.7}$$

where we have defined $K_{ab}$ as follows.

$$K_{ab} = R_{\mu\nu\rho\sigma} e^\mu_{(a)} k^\nu k^\rho e^\sigma_{(b)} \tag{1.6.8}$$

Now, we substitute the expression (1.5.27) in the case $N = 4$.

$$K_{ab} = -\frac{1}{2}R_{\rho\sigma}k^\sigma k^\rho \delta_{ab} + \sum_b C_{\mu\nu\rho\sigma}e^\mu_{(a)}k^\nu k^\rho e^\sigma_{(b)}\ell_b \qquad (1.6.9)$$

We now express the change in the cross-section when the ray flux advances by a small amount $d\lambda$ from the parameter $\lambda$ by the four-component quantity $A_{ab}$ defined by the following equation.

$$\frac{d\ell_a}{d\lambda} = A_{ab}\ell_b \qquad (1.6.10)$$

Since $A_{ab}$ can be regarded as a second-order tensor for rotation in a 2-dimensional plane perpendicular to the traveling direction of the ray bundle, we can decompose this tensor into three parts: trace part, symmetric trace free part and antisymmetric part. we write the decomposition as follows,

$$A_{ab} = \theta\delta_{ab} + \sigma_{ab} + \omega_{ab} \qquad (1.6.11)$$

where

$$\sigma_{ab} = \sigma_{ba}, \quad \sigma_{11} + \sigma_{22} = 0, \quad \omega_{ab} = -\omega_{ba} \qquad (1.6.12)$$

Changes due to the quantity $\theta$ change only the length without changing the direction of the vector $\ell_a$, so it represents the effect of changing the area of the ray bundle without changing its cross-sectional shape, and is called convergence. The quantity $\sigma_{ab}$ represents the effect of distorting the shape without changing the area since the trace is 0, and is called shear. Since $\omega_{ab}$ is antisymmetric, it represents the rotation of the cross-sectional area.

Since $\vec{\xi}$ is a vector connecting neighboring values of the same $\lambda$, $\vec{\xi}$, $\vec{k}$ are both expressed as partial differentials with respect to their parameters. Therefore, the relationship $[\vec{\xi}, \vec{k}] = 0$ holds, and the following equation is derived.

$$\frac{d\ell_a}{d\lambda} = \sum_b (\nabla_\nu k_\mu)\, e^\mu_{(a)} e^\nu_{(b)}\ell_b$$

Using this relation, each deformation of cross-section may be expressed in terms of photon four momentum as follows.

$$\theta = \frac{1}{2}\nabla_\mu k^\mu, \tag{1.6.13}$$

$$\sigma_{ab} = \nabla_{(\nu} k_{\mu)} e^\mu_{(a)} e^\nu_{(b)} - \frac{1}{2}\delta_{ab}(\nabla_\mu k^\mu), \tag{1.6.14}$$

$$\omega_{ab} = \nabla_{[\nu} k_{\mu]} e^\mu_{(a)} e^\nu_{(b)} \tag{1.6.15}$$

The geometric meanings of the Ricci and Weyl tensors become clear by expressing the expression for the geodesic deviation equation as a propagation equation for these quantities.

Differentiate equation (1.6.10) again by $\lambda$, we have

$$
\begin{aligned}
\frac{d^2\ell_a}{d\lambda^2} &= \sum_{b=1}^2 (\theta\delta_{ab} + \sigma_{ab} + \omega_{ab}) \frac{d\ell_b}{d\lambda} + \sum_{b=1}^2 \left(\frac{d\theta}{d\lambda}\delta_{ab} + \frac{d\sigma_{ab}}{d\lambda} + \frac{d\omega_{ab}}{d\lambda}\right)\ell_b \\
&= \sum_{b=1}^2 \left[\left(\theta^2 + \sigma^2 - \omega^2 + \frac{d\theta}{d\lambda}\right)\delta_{ab} + \frac{d\sigma_{ab}}{d\lambda} + \frac{d\omega}{d\lambda} + 2\theta\omega_{ab}\right]\ell_b
\end{aligned}
\tag{1.6.16}
$$

where we set the components of $\sigma_{ab}$, $\omega_{ab}$ as follows.

$$\sigma_{ab} = \begin{pmatrix} \sigma_1 & \sigma_2 \\ \sigma_2 & -\sigma_1 \end{pmatrix}, \quad \sigma^2 = (\sigma_1)^2 + (\sigma_2)^2, \quad \omega_{ab} = \begin{pmatrix} 0 & \omega \\ -\omega & 0 \end{pmatrix}$$

By comparing the equation (1.6.16) and the equation (1.6.7) the propagation equations for $\theta$, $\sigma_{ab}$, $\omega_{ab}$ are obtained as follows.

$$\frac{d\theta}{d\lambda} + \theta^2 + \sigma^2 - \omega^2 = -\frac{1}{2}R_{\mu\nu}k^\mu k^\nu \tag{1.6.17}$$

$$\frac{d\sigma_{ab}}{d\lambda} + 2\theta\sigma_{ab} = C_{\mu\nu\rho\sigma}e^\mu_{(a)}k^\nu k^\rho e^\sigma_{(b)} \tag{1.6.18}$$

$$\frac{d\omega_{ab}}{d\lambda} + 2\theta\omega_{ab} = 0 \tag{1.6.19}$$

The equation for the rotation $\omega_{ab}$ is linear in $\omega_{ab}$ and is homogeneous, so if you set it to 0 initially, it will always be 0 from then on, so we will not consider it anymore. From these equations, if we consider $\theta$, $\sigma$ as infinitesimal quantities and ignore their second and higher orders, then the Ricci tensor changes the cross-sectional area without changing the shape of the ray bundle, and the Weyl tensor does not change the area, but distorts the shape. Also, from the Einstein equation, the Ricci tensor is generated by the presence of matter, so changes in the cross-sectional area of the ray bundle occur as the ray propagates through the material.

Weak gravitational lensing, which will be explained later, is a phenomenon in which the shape of a tiny, distant galaxy is slightly deformed by a gravitational source such as a galaxy cluster. The deformation consists of the effect of expanding the size of the background galaxy and brightening it by the Ricci tensor (matter) and the effect of distorting its shape by the Weyl tensor.

## 1.7. Einstein Equation

Let us derive the gravitational field equation. For this purpose, we consider a fluid with energy density $\rho$. As seen above, the energy density is expressed by the energy momentum tensor as follows.

$$\rho = T_{\mu\nu}u^{\mu}u^{\nu} \tag{1.7.1}$$

Since this quantity is the right-hand side of Poisson equation, we use the energy-momentum tensor as the gravitational source of the field equation. Furthermore, we require the field equation is invariant under general coordinate transformations. Thus, we can naturally expect that there is a second-rank tensor $E^{\mu\nu}$ which balances the energy-momentum tensor.

$$E^{\mu\nu} = \kappa T^{\mu\nu} \tag{1.7.2}$$

with $\kappa$ a constant.

We also require that the energy-momentum tensor satisfies the conservation equation.

$$\nabla_\nu T^{\mu\nu} = 0 \tag{1.7.3}$$

This indicates that the tensor $E^{\mu\nu}$ should satisfies the same conservation

$$\nabla_\nu E^{\mu\nu} = 0 \tag{1.7.4}$$

We have already seen that Einstein tensor (1.5.37) satisfies the same equation, namely the contracted Bianchi identity (1.5.35). Thus, we conclude that the required equation takes the following form.

$$G^{\mu\nu} = \kappa T^{\mu\nu} \tag{1.7.5}$$

The constant $\kappa$ is determined such that the above equation reduces to Poisson equation in Newtonian limit. The result turns out to be $\kappa = 8\pi G$. Finally, we obtain the following field equation called Einstein equation.

$$G_{\mu\nu} = 8\pi G T_{\mu\nu} \tag{1.7.6}$$

In the case of vacuum, spacetime satisfies the vacuum Einstein equation.time

$$R_{\mu\nu} = 0 \tag{1.7.7}$$

As mentioned above, a Weyl tensor may exist even if the Ricci tensor vanishes identically, so this equation does not necessarily mean that spacetime is flat. In fact, there are vacuum solutions that are astronomically interesting, such as black holes and gravitational waves.

## Cosmological constant

As can be seen from the derivation of the Einstein equation above, there is a degree of freedom to add a term proportional to $g_{\mu\nu}$ in the left side. Not only is there no theoretical reason why this term is not 0, but there is also observational evidence that positively supports the existence of this term in cosmology, so that the following form is sometimes referred to as the Einstein equation.

$$G_{\mu\nu} + \Lambda g_{\mu\nu} = 8\pi G T_{\mu\nu} \qquad (1.7.8)$$

The second term on the left side is called the cosmological term, and the constant $\Lambda$ is called the cosmological constant. This term was introduced by Einstein, who believed that the universe was static before the discovery of the expansion of the universe, as a repulsive force (corresponding to $\Lambda > 0$) that balances gravity to create a static universe.

You can also include the cosmological term in the energy-momentum tensor on the right-hand side. To do this, we can define a stress energy tensor of a cosmological constant as follows.

$$T^{\Lambda}_{\mu\nu} = -\frac{\Lambda}{8\pi G} \qquad (1.7.9)$$

This form of the energy-momentum tensor is regarded to express a perfect fluid with the following energy density and pressure.

$$\rho_{\Lambda} = \frac{\Lambda}{8\pi G}, \quad P_{\Lambda} = -\rho_{\Lambda} \qquad (1.7.10)$$

The relationship between density and pressure $P = w\rho$ is called the equation of state, and $w$ is called the equation of state parameter. The cosmological constant is $w = -1$. As we will see later, the cosmological constant plays an important role in modern cosmology.

## 1.8.  Linear Approximation

Einstein's equation is a nonlinear partial differential equations for a metric tensor, and it is extremely difficult to solve analytically except in space-time with a high degree of symmetry, such as spherical symmetry or axial symmetry. Currently, much attention has been attracted in violent situations such as black hole mergers and neutron star mergers. Methods for solving Einstein's equations using numerical calculations have been developed. However, there are many interesting situations where gravity is weak and spacetime can be treated as slightly perturbed from flat Minkowski space-time in an appropriate coordinate system, as shown below. Furthermore, even for a strong gravitational source, the approximation considered here is not only not a bad approximation in a region that is more than several times the surface radius of the source, but also very useful for gaining physical intuition about general relativistic gravity.

We consider here the following form of the line element.

$$ds^2 = (\eta_{\mu\nu} + h_{\mu\nu})dx^\mu dx^\nu \tag{1.8.1}$$

where $h_{\mu\nu}$ is infinitesimally small quantity and can be treated as a tensor in Minkowski spacetime. The approximation that ignored the second order in $h_{\mu\nu}$ is called the linear approximation. The linear approximation is the basis for the description of gravitational lenses and gravitational waves.

### 1.8.1. *Linearized Einstein equation*

Let us derive Einstein equation in the linearized approximation. Riemann tensor takes the following form in this approximation.

$$R^{(L)}_{\mu\nu\rho\sigma} = \frac{1}{2}\left(h_{\mu\sigma,\nu\rho} - h_{\mu\rho,\nu\sigma} + h_{\nu\rho,\mu\sigma} - h_{\nu\sigma,\mu\rho}\right) \tag{1.8.2}$$

Thus, the linearized Einstein tensor is give by

$$G^{(L)}_{\mu\nu} = \frac{1}{2}\left(h^\rho{}_{\mu,\nu\rho} - h_{,\mu\nu} + h^{,\rho}{}_{\nu,\mu\rho} - h_{\mu\nu,\rho}{}^{;\rho}\right) - \frac{1}{2}\eta_{\mu\nu}\left(h^{\rho\sigma}{}_{,\rho\sigma} - h^{,\rho}{}_{,\rho}\right) \tag{1.8.3}$$

Now, we define a new variable as follows.

$$\bar{h}_{\mu\nu} = h_{\mu\nu} - \frac{1}{2}\eta_{\mu\nu}\eta_{\mu\nu}h \tag{1.8.4}$$

Using he variable, the linearized Einstein tensor takes the following form.

$$G^{(L)}_{\mu\nu} = \frac{1}{2}\left(\bar{h}^\rho{}_{\mu,\nu\rho} - \bar{h}_{,\mu\nu} + \bar{h}^\rho{}_{\nu,\mu\rho} - \bar{h}_{\mu\nu,\rho}{}^{;\rho}\right) \tag{1.8.5}$$

**Harmonic condition**

We can further impose the following condition to our variable $\bar{h}_{\mu\nu}$.

$$\bar{h}^{\mu\nu}{}_{,\nu} = 0 \tag{1.8.6}$$

The condition is called the harmonic gauge condition or simply harmonic condition. The reason of the name is as follows. Let require the following equation to hold for the spacetime coordinates.

$$\eta^{\rho\sigma}\nabla_\rho\nabla_\sigma x^\mu = 0 \tag{1.8.7}$$

Now, we regard that the coordinates are scalar functions and calculate (1.8.17) to find[3]

$$\nabla_\rho\nabla^\rho x^\mu = \frac{1}{\sqrt{-g}}\frac{\partial\left(\sqrt{-g}\nabla^\rho x^\mu\right)}{\partial x^\rho} = \frac{1}{\sqrt{-g}}\frac{\partial\left(\sqrt{-g}g^{\mu\rho}\right)}{\partial x^\rho} = 0 \tag{1.8.8}$$

Thus the condition becomes the following condition for the metric tensor.

$$\frac{\partial\left(\sqrt{-g}g^{\mu\rho}\right)}{\partial x^\rho} = 0 \tag{1.8.9}$$

It is easy to show that this condition agrees with condition (1.8.6) in the linearized order in $h$. In the case of a positive definite metric, a function satisfying the equation (1.8.7) is called as a harmonic function. That is the reason the condition (1.8.7) is called harmonic.

**Exercise.** An infinitesimal coordinate transformation is written as follows.

$$x^\mu \to x^{\mu'} = x^\mu + \xi^\mu \tag{1.8.10}$$

Find the condition for function $\xi^\mu(x)$ that transform an arbitrary coordinate system to the coordinate system satisfying the harmonic condition.

---

[3]Infinitesimal displacements $dx^\mu$ of the coordinate $x^\mu$ is a vector under the general coordinate transformation, but the coordinate itself is not a vector.

The linearized Einstein equation takes the following simple form in the harmonic coordinate condition.

$$\eta^{\rho\sigma}\partial_\rho\partial_\sigma \bar{h}^{\mu\nu} = -16\pi G T^{\mu\nu} \qquad (1.8.11)$$

Since the left-hand side of this equation satisfies the harmonic condition, the energy-momentum tensor in the right-hand side satisfies the conservation law in the flat spacetime.

$$T^{\mu\nu}_{\,\,,\nu} = 0 \qquad (1.8.12)$$

Therefore, linearized Einstein equation (1.8.11) can expresses the lowest order gravitational field created by a source, but it is not able to describe the interaction between matter and gravitational field. This indicates that the linearized theory is not able to treat the change of motion by the energy loss due to the emission of gravitational waves.

## Gravitational field around a static body

The difference between general relativistic gravity and Newtonian gravity is noticeable in strong gravity such as that of a black hole and in the collision of compact stars, but in many astronomical situations, such as the motion of planets around the sun or the gravitational lensing effect described below, gravity is weak and its time dependence can be ignored. We will derive gravitational field around a body in such a situation.

Since we ignore the time dependence, the linearized Einstein equation in the harmonic coordinate can be written as follows.

$$\Delta \bar{h}^{\mu\nu} = -16\pi G T^{\mu\nu} \qquad (1.8.13)$$

This equation has the following solution under the condition that spacetime approaches flat at infinity

$$\bar{h}^{\mu\nu}(x^k) = 4G \int d^3 y \, \frac{T^{\mu\nu}(y^k)}{|x^k - y^k|} \qquad (1.8.14)$$

We are interested in the region outside the source of gravity. Thus, we evaluate the integral above under the condition $r = |\boldsymbol{x}| \gg |\boldsymbol{y}|$ where the spatial origin $\boldsymbol{x} = \boldsymbol{0}$ is taken to be inside the source.

$$\bar{h}^{\mu\nu}(\boldsymbol{x}) = \frac{4G}{r} \int d^3y\, T^{\mu\nu}(y^k) + O\left(\frac{1}{r^2}\right) \tag{1.8.15}$$

Four velocity of the source $\vec{u}$ is calculated by the normalization condition $g_{\mu\nu}u^\mu u^\nu = -1$ as

$$u^\mu = \frac{1}{\sqrt{1 - v^2 - h_{00}}} \tag{1.8.16}$$

We choose the center of mass as the spatial origin and assume that the source as a whole is at rest. Then we have $\rho = T^{00}$ since $h_{00}$ is the first order quantity.

$$M = \int d^3y\, T^{00} \tag{1.8.17}$$

On the other hand, $T^{\mu\nu}{}_{,\nu} = 0$ gives

$$\frac{d}{dt} \int d^3y\, T^{i0} = -\int d^3y\, T^{ij}{}_{,j} = \oint d^2S_j\, T^{ij} = 0 \tag{1.8.18}$$

Thus, the following three momomentum is a constant.

$$P^i = \int d^3y\, T^{i0} \tag{1.8.19}$$

Since we are in the center of mass frame, $T^{0i} = \rho v^i = 0$. Using the conservation law twice, we can derive the following relation.

$$\ddot{I}^{ij}(t) = \int d^3y\, T^{ij}(y^k, t) \tag{1.8.20}$$

where $I^{ij}$ is the quadrupole moment of the source defined as

$$I^{ij}(t) = \int d^3y\, y^i y^j T^{00}(y^k, t) \tag{1.8.21}$$

Since we here consider a static source, the time derivative of the quadrupole moment vanishes. Finally, we obtain the following gravitational field around a static source applicable well beyond the size of the source.

$$\bar{h}^{00}(\boldsymbol{x}) = \frac{4GM}{r} + O\left(\frac{1}{r^2}\right), \quad \bar{h}^{0i} = \bar{h}^{ij} = O\left(\frac{1}{r^2}\right) \quad (1.8.22)$$

The line element in this region is expressed as follows.

$$ds^2 = -\left(1 - \frac{2GM}{r}\right)dt^2 + \left(1 + \frac{2GM}{r}\right)(dx^2 + dy^2 + dz^2) \quad (1.8.23)$$

Newtonian gravity does not assume any curvature of space, so the coefficient of the second term on the right side in the above equation is 1. As will be explained later, this difference is the reason why the angle of light bending due to the gravitational field in general relativity is twice the bending angle of Newtonian gravity.

# Chapter 2

# Brief Introduction to Cosmology

We now know from a variety of observations that the universe began expanding approximately 13.8 billion years ago in an ultra-high temperature, ultra-dense state known as the Big Bang, and has continued to expand, and is now expanding at an accelerated rate. Furthermore, space is almost flat, and the material made of protons, neutrons, and electrons (called baryonic matter) that makes up stars and our bodies has only 5% of the energy density of the present universe. We also know that the remaining 95% is unidentified energies called dark matter and dark energy.

Dark matter does not interact with electromagnetic waves at all, but its gravity creates structures such as galaxies, galaxy clusters, and superclusters from small fluctuations in the early universe, and dark energy plays an important role in counteracting gravity and accelerating the expansion of the universe. One of the purposes of observational cosmology is to explore the true nature of dark matter and dark energy and to clarify the details of structure formation in the universe.

Observations such as cosmic microwave background radiation (CMB) and the large scale 3-dimensional distribution of galaxies have shown that our universe is globally homogeneous and isotropic in a good approximation, and as far as we are concerned with the temporal evolution of the entire universe and the global geometry, it is useful to assume that the space is homogeneous and isotropic.

However, in order to understand the formation of structures in the universe, it is necessary to consider deviations from homogeneity and isotropy. General relativity is indispensable to deal with such deviation. This chapter describes the basic observations and theoretical background of the observational cosmology.

## 2.1. Summary of Observations

Current cosmology is an empirical science based on many observed facts. Among these observations, here we will summarize the observations of the expansion of the universe, galaxy redshift survey, and cosmic microwave background radiation, which will be necessary for later understanding.

### 2.1.1. *Cosmic Expansion*

The first step in the development of the modern universe was the discovery by American astronomer Leavitt of the relationship between the period and luminosity of Cepheid variable stars in the 1910s. The discovery of the relation that the longer the period of variation, the brighter the average luminosity made it possible to estimate the distances of distant galaxies. A celestial object whose absolute brightness can be accurately estimated is called a standard candle. Absolute magnitude $M$ is defined as the apparent magnitude $m$ when the celestial body is placed at a distance of $10\,\mathrm{pc}$.[1]

$$m - M = 5\log\left(\frac{d}{10\,\mathrm{pc}}\right) \qquad (2.1.1)$$

The distance defined by this relationship is precisely called the luminosity distance.

---

[1]Actual observations are performed in specific wavelength bands. The magnitude value depends on the observed wavelength band.

Estimates of distances at that time were not accurate because the existence of two types of Cepheid variables were not recognized, but they were still sufficient to make us realize that the Andromeda galaxy was an object outside our Milky Way galaxy. Furthermore, using this relationship, E. Hubble and his collabrator discovered that galaxies tend to move at a receding velocity $V$ that is proportional to their distance $d$.

$$V = H_0 d \tag{2.1.2}$$

This is called the Hubble–Lemaitre law. G. Lemaitre was a Belgian physicist who proposed this law ahead of Hubble. $H_0$ is called the (current) Hubble parameter.

Our galaxy cannot be the center of the universe, and in order for this law to hold true even when observed from all galaxies, it is necessary to understand the Hubble–Lemaître law as an expansion of space. In other words, suppose that any two galaxies are fixed in space and their distance $\ell$ is constant, but space itself is expanding according to a certain function $a(t)$. Then, the physical distance can be expressed as a function of time as follows.

$$D(t) = a(t)\ell \tag{2.1.3}$$

If we differentiate this equation with time, we obtain the Hubble–Lemaître law.

$$V = \frac{dD(t)}{dt} = \frac{da}{dt}\ell = \frac{\dot{a}}{a}D(t) \tag{2.1.4}$$

In this way, the Hubble–Lemaître law can be interpreted as the expansion of space. Here, the function $a$ that represents expansion is called the scale factor, and the Hubble parameter is expressed using the scale factor as follows.

$$H(t) = \frac{\dot{a}}{a} \tag{2.1.5}$$

Therefore, the Hubble parameter is a function of time, and $H_0$ is its current value. Its value is estimated to be around $70\,\mathrm{km s^{-1} Mpc^{-1}}$. This unit means that the relative velocity of two galaxies 1 Mpc apart is $100\,\mathrm{km/s}$.

Since the Hubble parameter $H$ has the inverse dimension of time, the inverse provides a measure of cosmic time up to that time. This is called the Hubble time, and the value multiplied by the speed of light is called the Hubble radius or the Hubble horizon radius. The Hubble horizon radius at any given time provides a measure of the causally influential scale at that time. The Hubble time for the current Hubble parameters is

$$t_{H,0} = \frac{1}{H_0} = 9.8 \times 10^9 h^{-1} \text{yr} \qquad (2.1.6)$$

where $h$ is the normalized Hubble parameter defined as follows.

$$H_0 = 100 h \, \text{kms}^{-1} \text{Mpc}^{-1} \qquad (2.1.7)$$

**Distance ladder**

Figure 2.1 shows the relationship between distance and recession velocity using Cepheid variables obtained by the Hubble Space Telescope. The period-luminosity relationship for Cepheid variables becomes practical only when the absolute distance of a variable star with a certain period is known. To do this, it is necessary to observe relatively nearby Cepheid variables in our galaxy whose distances can be measured from the measurment of the annual parallax (triangulation using the distance between the sun and the earth as the base). Currently, the annual parallax is measured with great precision by the Hubble Space Telescope (HST) and the positioning astronomy satellite Gaia. HST can accurately measure distances of about 10,000 light years, and Gaia can accurately measure distances of up to about 30,000 light years. For example, the following period-magnitude relationship has been obtained from distance measurements of several dozen Cepheid variable stars using HST. (Riess *et al.* *ApJ*, 826, 56, 2015).

$$M_H = -3.26 \left( \log_{10} P(\text{day}) - 1 \right) - 5.93 \qquad (2.1.8)$$

where $M_H$ is the absolute magnitude measured by H-band (the wavelength is around $1.63 \, \mu\text{m}$).

Hubble Diagram for Cepheids (flow−corrected)

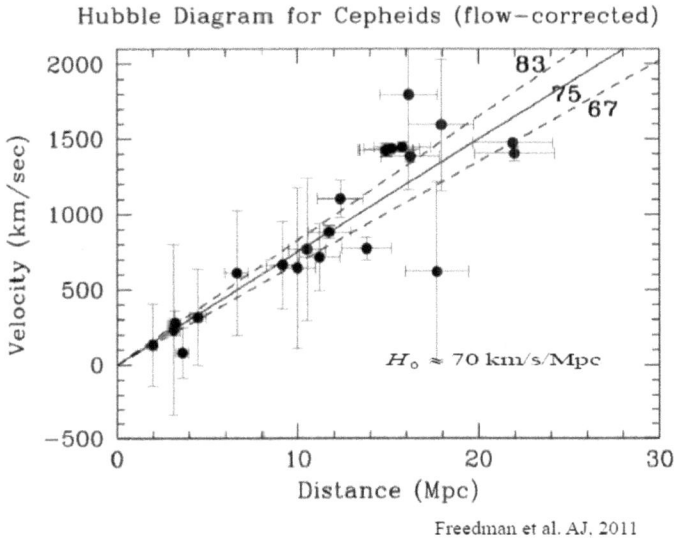

Freedman et al. AJ. 2011

Fig. 2.1. Hubble space telescope observations of Hubble–Lematre parameter by Cephaid variables.

As can be seen from this relation, the absolute magnitude of a Cepheid variable, even the brightest, is about −8, which is about 100,000 times brighter than the absolute magnitude of the Sun, which is +4..82. Even so, it is impossible to resolve galaxies more than 30 Mpc into stars, even using a space telescope like HST. Therefore, in order to show that Hubble's law holds true for galaxies that are more distant, we need a brighter standard candle. To date, several candidates for standard candles are known, such as the relationship between the rotation speed and brightness of disk galaxies, but the most reliable of these is the explosion phenomenon of white dwarfs called type Ia supernovae. The typical brightness of this supernova is about −19, which is comparable to the brightness of a small galaxy. Therefore, the distance to distant galaxies exceeding 1 Gpc can be estimated using the supernova. It has been shown that distance measurements for relatively nearby galaxies using Cepheid variable stars and type Ia supernovae are consistent. The method of measuring distant distances by starting from triangulation and

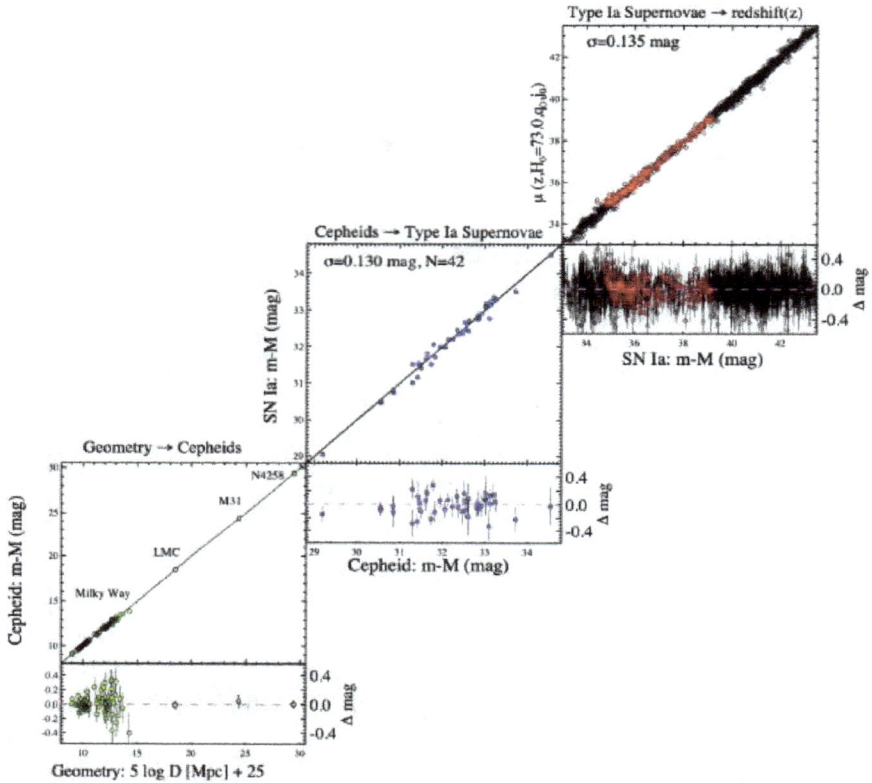

Fig. 2.2. Distance ladder. (A.G. Riess *et al.*, The Astrophysical jouranal Letters, 934, L7, 2022).

connecting different standard light sources without contradiction is called the distance ladder method (see Fig. 2.2). The accuracy of the conventional distance ladder has been verified by the James Webb Space Telescope (JWST), which was launched in 2021.

Figure 2.3 shows the relationship between the apparent brightness and redshift of a large number of type Ia supernovae obtained in this way. The vertical axis is the difference between the apparent magnitude and the absolute magnitude, which is called the distance modulus, and represents the distance using the equation (2.1.1).

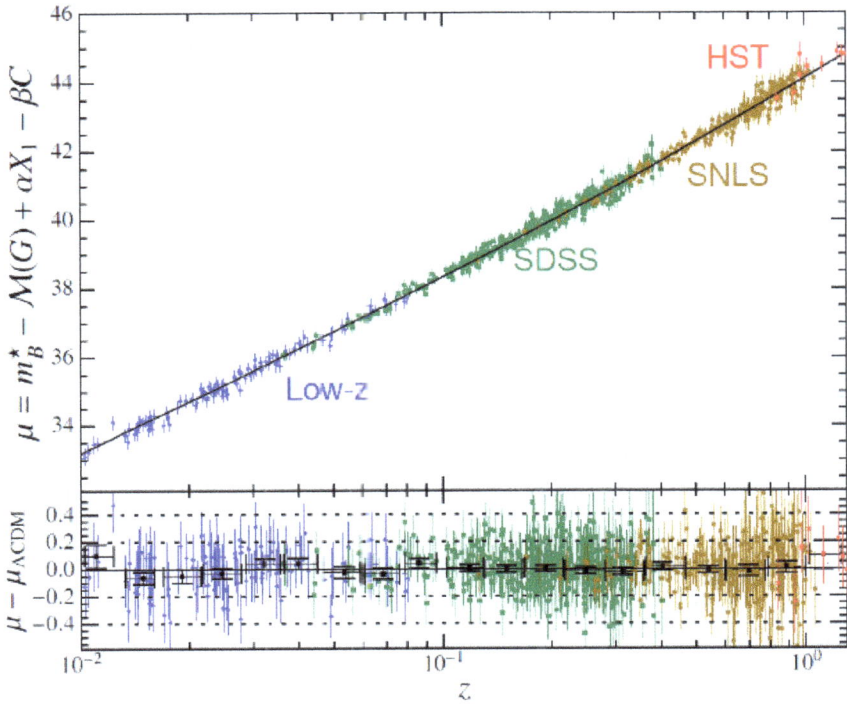

Betoule et.al, 2014 A&A

Fig. 2.3. The apparent magnitude-redshift (m-z) relation observed by Type Ia supernovae.

The redshift $z$ on the horizontal axis is defined by the following equation.

$$z = \frac{\lambda_{\rm obs} - \lambda_{\rm em}}{\lambda_{\rm em}} \qquad (2.1.9)$$

where $\lambda_{\rm em}$ and $\lambda_{\rm obs}$ are the wavelength emitted in the rest frame of distant galaxy and the wavelength at which it was received, respectively. The recession velocity of distant galaxies is due to the expansion of space, and is not due to the simple Doppler effect, but when the redshift is sufficiently small compared to 1, the relationship

$v = cz$ holds ($c$ is the speed of light). If $z$ becomes larger than about 0.1, other consideration is required as described later, but for now, it can be considered that it simply corresponds to speed. Therefore, Figure 2.3 is equivalent to the Hubble–Lemaître law.

Current value for the Hubble parameter observed by m-z relation of type Ia supernvae is estimated to be as follows. (Riess *et al.* 2019)

$$H_0 \simeq 74.03 \pm 1.42\,\text{km/sec/Mpc} \qquad (2.1.10)$$

### Discovery of accelerated expansion and dark energy

Gravity determines the rate of expansion of the universe, and since gravity is an attractive force, the rate of expansion is expected to decrease in time (decelerating expansion). However, at the end of the 20th century, there was a discovery that contradicted this prediction. It was shown by two independent groups that the $m - z$ relation obtained by about 500 Type Ia supernpvae indicates that the current expansion of the universe is accelerating.

It is assumed that the cause of this accelerated expansion is the existence of an unknown energy called dark energy that exerts a repulsive force. The cosmological constant that Einstein introduced to create a static universe is a leading candidate of dark energy. However, the true nature of dark energy is currently completely unknown. Rather than assuming such an unknown energy source, the possibility that the law of gravity differs from the general theory of relativity on a cosmological distance scale (modified gravity theory) is also being discussed. Since there have been no experiments or observations that actively suggest the modification of general theory of relativity, we will not discuss this possibility anymore. The investigation of the dark energy is one of the most important themes in modern cosmology, and it is expected that the nature of dark energy will be elucidated through observations of baryon acoustic oscilation discussed below and cosmic shear using weak lensing discussed in the next chapter.

## 2.1.2. *Discovery of cosmic microwave background (CMB) and the temperature anisotropy*

In 1964, A. Penzias and R. Wilson discovered microwaves that fill the universe. This is a remnant of an ultra-dense thermal equilibrium state, and is decisive evidence for the Big Bang theory. Later, precise observations by NASA's CMB observation satellite revealed that this microwave has almost perfect Planck spectrum with an absolute temperature of 2.725 K, and confirmed that the temperature distribution was isotropic with an accuracy of $10^{-5}$. The discovery of small diviation of isotropy made cosmology an exact science.

### The origin of CMB

The fact that the CMB has a Planck distribution even though the current universe is not in thermal equilibrium is evidence that the early universe was in thermal equilibrium. In a state of thermal equilibrium, the energy density of radiation is proportional to the fourth power of temperature and inversely proportional to the fourth power of the scale factor (the third power is volume dependence and another power comes from the wavelength dependent of energy of a photon), so the temperature of radiation is inversely proportional to the factor. Therefore, the earlier the universe is, the higher the temperature, and matter is broken down into its more basic constituent elements. In this book, we consider the universe after the temperature drops to below tens of thousands of degrees. Approximately 50,000 years after the Big Bang, the temperature of the universe reached approximately 10,000 degrees Celsius, and the universe was filled with a fluid in which completely ionized plasma consisting of electrons, protons, and helium nuclei and a large number of photons were strongly coupled by Thomson scattering and behaved as a one-component fluid.[2] This fluid is called a photon-baryon fluid.

---

[2]Neutrinos and dark matter do not interact with photons and baryons, so they are not relevant to the discussion here.

How much baryons are contained in this fluid plays an important role in the discussion below. This is characterized by the following parameter.

$$\eta \equiv \frac{n_b}{n_\gamma} = (5.9 - 6.4) \times 10^{-10}$$

The value is observed by theory and observation of light element synthesis in the early universe which we will not discuss in this book. Also it is worth mentioned that the value is related with the asymmetry between matter and antimatter, which is still not understood.

Even if the temperature of the universe drops to around $13.6\,\text{eV} \simeq 15{,}000\,\text{K}$, which is the ionization energy of hydrogen, there will still be a large number of photons in the high frequency part of the Planck distribution. The plasma state is maintained even at lower temperatures. When the temperature of the universe drops to about 3800 degrees and there are fewer photons with energy greater than hydrogen's ionization energy of $13.6\,\text{eV}$, protons capture electrons and hydrogen atoms begin to form.[3] This is called recombination. Then, about 380,000 years after the Big Bang, when the temperature of the universe dropped to about $3000\,\text{K}$, the number of free electrons deceases enough to stop interacting with baryonic matter and begin to travel freely in a straight line. This phenomenon occurred at redshift $z \simeq 1090$ and is called the decoupling of the universe or the clearing up of the universe. The observed CMB is red-shifted photons scattered lastly at the decoupling. Sometimes we use the the time of decoupling as last scattering surface.[4]

---

[3]Helium atomic nucleus has already captured two electrons and has become a neutral helium atom at a temperature about $7000\,\text{K}$.

[4]After that, due to the birth of protostars, most neutral atoms Ionized. This is called reionization in the universe.

## CMB temperature fluctuations and their origin

In 1989, the CMB observation satellite COBE launched by NASA discovered that the CMB has a perfect Planckian distribution and fluctuations of temperature of the order of of $10^{-5}$. Temperature fluctuation is defined as the difference in temperature in two directions $n$, $n'$ separated by an angle $\theta$ as follows.

$$\Theta(\theta) = \frac{T(n) - T(n')}{T_0} \tag{2.1.11}$$

Here, $T_0 = 2.726\,\mathrm{K}$ is the average temperature of the CMB. Figure 2.4 shows that the temperature map observed by Planck satellite launched 2009 by European Space agency. with a temperature resolution of the order of $10^{-6}$ at all angular resolution greater than 10 arcminutes.

Density fluctuations of baryons in the photon-baryon fluid cannot grow due to strong radiation pressure and oscillate and propagate through the fluid as sound waves (compression waves). The temperature fluctuation of the CMB is the result of this sound wave vibration pattern remaining on the last scattering surface. Its fundamental wavelength is the distance that a sound wave propagated from the Big Bang until the time of last scattering. The distance is called

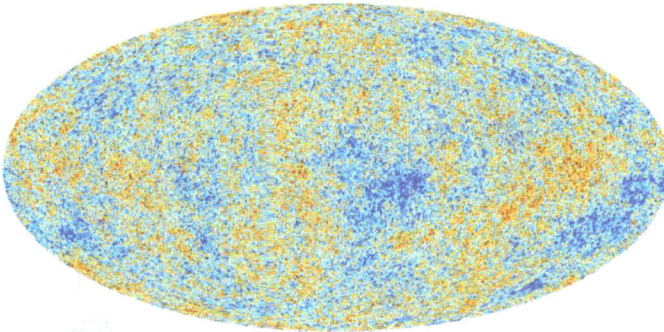

Fig. 2.4. Temperature map observed by PLanck.

the acoustic horizon radius. The speed of sound is determined by the ratio of the number density of photons and baryonic particles, and the time of last scattering is determined by the radiation density and the matter density (dark matter, baryonic matter). As we will see later, the recombination time and the speed of sound do not depend on the global geometry of the universe or the cosmological constant. Therefore, the radiation density, matter density and the ratio of the baryon number to the photon number can be determined from the observation of temperature fluctuations, regardless of the spatial curvature or the cosmological constant. Furthermore, the physical distance of the acoustic horizon is measured at a certain angle on the celestial sphere, and that angle depends on the global geometry of the universe. In this way, observation of temperature fluctuations in the CMB provides a lot of information about cosmological parameters.

The remnants of this sound wave also brings about a periodic distribution in the observed spatial distribution of galaxies with period of about 147 Mpc. This periodic pattern is called baryon acoustic oscillations (BAO), and will be discussed later. The physical scale of the acoustic horizon and BAO are regarded as standard rulers and play an important role in measuring cosmological parameters. We will discuss on this more detail later.

Figure 2.5 shows the power spectrum of the CMB temperature flcutation obtained by Planck satellite. The power spectrum is the power of the fluctuation in each scale of perturbation and is defined as follows. We first expand temperature fluctuations using spherical harmonics as follows.

$$\Theta(\boldsymbol{n})_{\mathrm{obs}} = \sum_{\ell=1}^{\infty} \sum_{m=-\ell}^{\ell} a_{\ell m} Y_{\ell m}(\vec{n}) \qquad (2.1.12)$$

The expansion coefficient $a_{\ell m}$ represents the amplitude of each $\ell$ pattern. $\ell = 1$ represents the dipole distribution, which gives information about the motion of the Earth relative to a system in which the CMB appears isotropic. We will not discuss this anisotropy

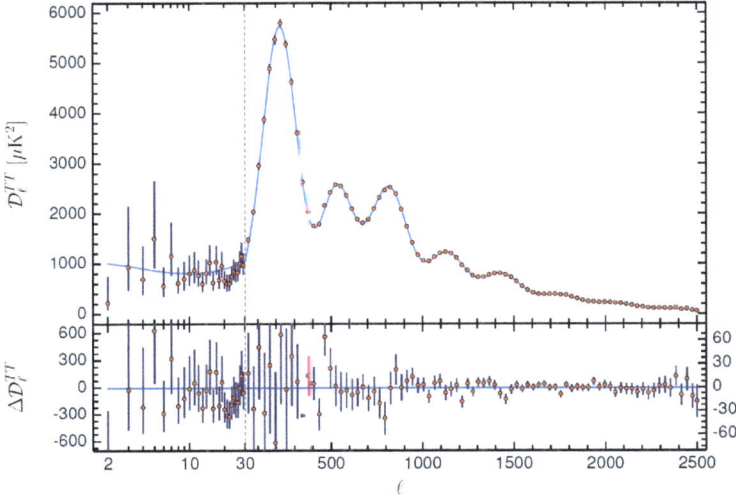

Fig. 2.5. CMB power spectrum observed by Planck satellite.

in this book. The temperature fluctuations in CMB of primordial origin are imprinted in the modes larger than $\ell = 2$ (quadrupole distribution). The larger $\ell$ is, the smaller the scale of fluctuation is. It is proportial to the inverse of angle $\theta$ on the celestial sphere, and since the dipole distribution cf $\ell = 1$ corresponds to 180 degrees, the angle has the following relationship.

$$\ell \simeq \frac{180^o}{\theta^o} \tag{2.1.13}$$

The temperature fluctuations have random values in each direction, but they are statistically isotropic and can be characterized by the following ensemble average.

$$\langle a_{\ell m} a^*_{\ell' m'} \rangle = \delta_{\ell\ell'} \delta_{mm'} C_\ell \tag{2.1.14}$$

This $C_\ell$ is called the power spectrum of temperature fluctuation. If the temperature fluctuation is Gaussian, the statistical properties are determined only by the power spectrum. Observationally, non-Gaussian components such as three-point correlation of fluctuation has not been observed yet. Using this power spectrum, the root mean square of temperature fluctuation can be expressed as follows:

$$\langle |\Theta_{\text{obs}}|^2 \rangle = \sum_{\ell,\ell',m,m'} \langle a_{\ell m} a^*_{\ell' m'} \rangle Y_{\ell m}(\boldsymbol{n}) Y_{\ell' m'}(\boldsymbol{n}) = \sum_{\ell} \frac{2\ell+1}{4\pi} C_\ell$$

(2.1.15)

For $\ell \gg 1$, the sum is approximated as

$$\sum_{\ell} \frac{2\ell+1}{4\pi} C_\ell \simeq \int d\ln \ell \frac{\ell(2\ell+1)}{4\pi} C_\ell \simeq \int d\ln \ell \frac{\ell(\ell+1)}{2\pi} C_\ell \quad (2.1.16)$$

Thus, $\ell(\ell+1)C_\ell/2\pi$ is the power per logarithmic interval in angular space, and this quantity is plotted in Figure 2.4. The appearance of several peaks is due to the sound waves propagating in the baryonic fluid mentioned above. Various cosmological parameters can be determined from the positions of the peaks of this fluctuation and the differences in height between each peak.

The peak appeared at $\ell \simeq 220$ in Fig. 2.6 is called the first acoustic peak which corresponds to the acoustic horizon at the last scattering surface, and is observed as.

$$\theta_H(z_{\text{LS}}) = 0^o.59643 \pm 0^o.00046$$

(2.1.17)

This angle mainly depends on the spatial curvature of the universe, but it also slightly depends on matter density and baryon density, so this information alone cannot completely determine the exact curvature. Taking other informations into consideration, the following strict limitation is obtained.

$$\Omega_{K,0} = -0.012 \pm 0.010$$

(2.1.18)

$\Omega_{K,0}$ is the current value of the dimensionless curvature parameter and its definition will be given below. Because of this limitation, the standard model assumes that the universe is flat.

Assuming a flat universe, the following values are obtained as the best fit of the cosmological parameters from Figure 2.4. (see section

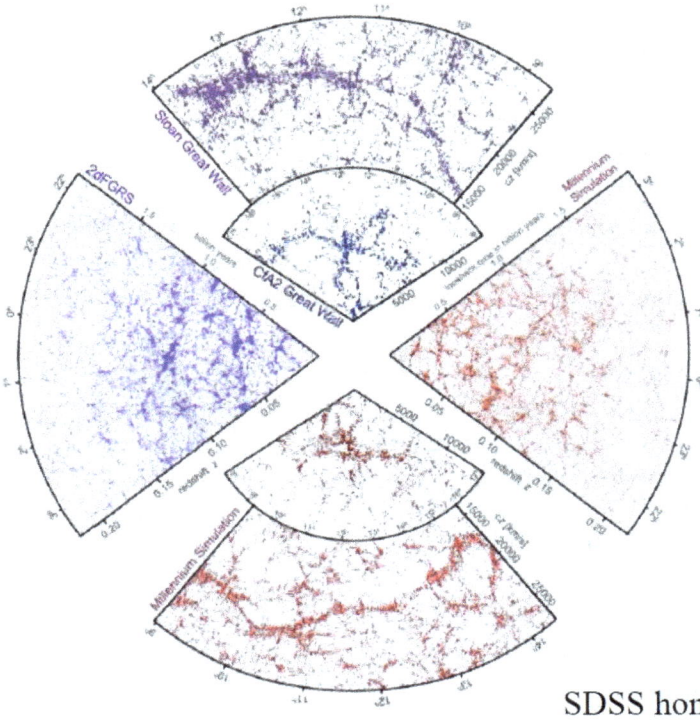

SDSS homepage

Fig. 2.6. Three-dimensional distribution of galaxies by CfA2 (1985–1995) and by SDSS (2000–).

below for meanings of symbols).

$$\Omega_{b,0}h^2 = 0.002226 \pm 0.00013,$$

$$\Omega_{c,0}h^2 = 0.1190 \pm 0.0011,$$

$$\Omega_{m,0} = 0.3092 \pm 0.0066,$$

$$H_0 = 67.66 \pm 0.49 \,\mathrm{km/s/Mpc},$$

$$\sigma_8 = 0.8113 \pm 0.0050,$$

$$t_0 = (13.7971 \pm 0.023) \times 10^{10} \,[\mathrm{yr}],$$

$$z_{\mathrm{dec}} = 1089.92 \pm 0.25,$$

$$z_{\mathrm{eq}} = 3402 \pm 26.$$

Here, the values up to $H_0$ are from the 2023 data, and the values after that are from the 2018 data.

## Hubble tension

You will note that two different values appear for the Hubble parameter. The value of the Hubble parameter calculated from the galaxy's recession velocity and the apparent brightness of the supernova is about $h \simeq 0.74$. On the other hand, CMB observations gives $h = 0.677$. Current cosmology is based on the observation that the global space of the universe is uniform and isotropic. Therefore, it is expected that the Hubble parameter, which represents the expansion rate of the universe, will have the same value no matter how it is measured, but this result clearly contradicts that expectation. This problem is called the Hubble tension.

It is not known whether the values of the Hubble parameters are actually different, or whether they appear to be different due to observational errors and will match in future more accurate observations.

Measurements using galaxies and supernovae actually measure distance and velocity, so they are direct measurements, whereas measurements using CMB are based on the uniformity and isotropy of space and the physical laws of the early universe. This is an indirect method of constructing a model, reproducing the power spectrum of the CMB based on it, and adjusting cosmological parameters to make it match the observed power spectrum. Since this method is based on indirect but simple assumptions of spatial geometry and clear physics, the error in Hubble parameter measurements is very small with $0.5 \, \mathrm{kms}^{-1}\mathrm{Mpc}^{-1}$. On the other hand, direct measurements are based on the assumption that Cepheid variables and Type Ia supernovae are the standard candles. These celestial bodies are not standard candles in the exact sense, but each Cepheid variable and Type Ia type supernova have their own individuality, and in order to consider them as standard candles, empirical corrections must be made to eliminate their individuality. Although some ambiguity exists in the direct method, there is no contradiction in the distance ladder used so far, and the reported error is also small $\pm 1\,\mathrm{kms}^{-1}\mathrm{Mps}^{-1}$ At the moment we donot know if the tension is real or will disappear in future.

### 2.1.3. *Galaxy redshift survey and matter power spectrum*

Most galaxies in the present universe belong to structures of various scales, such as galaxy groups, galaxy clusters, and superclusters. According to current understanding, the structures in the distribution of galaxies is thought to have been formed by the gravitational growth of small fluctuations in the matter density. The origin of the matter fictuation is thought to be quantum fluctuations during the inflationary expansion in the early universe.

Therefore, the spatial distribution of galaxies is determined by the initial matter density spectrum (the scale dependence of the density fluctuation) and the subsequent growth of the fluctuation determined by the balance between gravity and the expansion of the universe. The power spectrum of fluctuations is determined by the physics of the early universe, and gravity is determined by the amount of dark matter and its nature. Dark energy is also controlling the current expansion of the universe. Observations of the spatial distribution of galaxies thus provide important information not only on the structure formation but also on the early Universe, dark matter and dark energy.

In order to know the statistical properties of spatial distribution of galaxies the redshift information of huge number of galaxies in a wide regio of sky is required. but spectroscopic observations for distant galaxies require extremely long observation time. This is the reason why large-scale redshift surveys were not conducted until CCD became available for astronomical observation.

Large-scale redshift survey began in the 1980s, and the largest survey so far called Sloan Digital Sky Survey(SDSS) was jointly launched by the United States and Japan in 1999 and later Max-Planck Institute for Astrophysics (Germany) was joined. The survey covered a quarter of the entire sky and comprehensively observed celestial bodies up to 22 magnitude. Of these, the redshifts of about 1 million galaxies and about 100,000 quasars have been measured. This has revealed that the distribution of galaxies is not uniform, and that many galaxies exist as members of galaxy clusters on a scale of 2 to 3 Mpc, and that galaxy clusters are forming even larger scale superclusters. However, when these structures are averaged on the 150 Mpc scale, there is no large difference in the number density of galaxies, and in this sense, the matter distribution in the universe is considered to be globally uniform.

In order to extract various information from the spatial distribution of galaxies, it is necessary to quantify its distribution. For this purpose, the two-point correlation function of the galaxy or the power spectrum, which is its Fourier transform, is used. Define the number density distribution of galaxies as follows.

$$n_g(\boldsymbol{x}) = \bar{n}_g(t)\,(1 + \delta_g(\boldsymbol{x}, t)) \tag{2.1.19}$$

where $\bar{n}_g(t)$ is the mean number density. Then the 2-point correlation function and the power spectrum are defined as follows.

$$\xi_g(r) = \langle \delta_g(\boldsymbol{x})\delta_g(\boldsymbol{y})\rangle, \tag{2.1.20}$$

$$P_g(k) = \int d^3x\, e^{-i\boldsymbol{k}\cdot\boldsymbol{x}}\xi_g(r) = 4\pi \int_0^\infty dr\, r^2 \xi_g(r)\frac{\sin kr}{kr} \tag{2.1.21}$$

The two-point correlation function depends only on the distance between two points $r = |\boldsymbol{x} - \boldsymbol{y}|$ due to the isotropy of space. However, since most of the mass in the universe is due to dark matter, the fluctuations in the number density of galaxies $\delta_g$ do not directly represent the fluctuations in the mass density $\delta_m$. To know the mass distribution from the galaxy distribution, it is necessary to know the relationship between the galaxy distribution and the dark matter distribution. This relationship is called a bias, and it is very difficult to derive theoretically because it depends on the details of galaxy formation. The phenomenological assumption that brightness (galaxies) and mass (dark matter) are proportional is often used.

$$\delta_m = b\delta_b \tag{2.1.22}$$

This assumption is called the linear bias. This proportionality coefficient $b$ depends on the scale considered. From now on, we assume the linear bias and consider the matter density fluctuations including dark matter, as well as baryonic matter and its spectrum will be written as

$$\langle \hat{\delta}_m(\boldsymbol{k})\hat{\delta}_m^*(\boldsymbol{k}')\rangle = (2\pi)^3 \delta^{(3)}(\boldsymbol{k} - \boldsymbol{k}')P_m(k) \tag{2.1.23}$$

where $\delta^{(3)}$ is 3-dimensional Dirac delta function, and $\hat{\delta}$ is the Fourier component of the density fluctuation.

One can easily confirm that the Fourier transform of the correlation function $\xi(r) = \langle \delta_m(\boldsymbol{x})\delta_m(\boldsymbol{x} + \boldsymbol{r})\rangle$ is the power spectrum defined above.

$$P_m(k) = \int d^3 r\, e^{-i\boldsymbol{k}\cdot\boldsymbol{r}}\xi(r) \tag{2.1.24}$$

Figure 2.5 shows the power spectra of matter density fluctuations in the current Universe obtained by various observations.

## Dispersion of mass fluctuation

One can understand the meaning of the matter power spectrum as the dispersion of mass fluctuation. We write the mass contained in a region with volume $V = R^3$ as

$$M_V = \int_V d^3 x\, \rho_m(\boldsymbol{x}) = \rho_m \int_V d^3 x\,(1 + \delta_m(\boldsymbol{x})) = \bar{M}_V + \delta M \tag{2.1.25}$$

where $\bar{M}_V = \bar{\rho}_m V$ is the averaged mass contained in the volume $V$. Thus, the mass fluctuation can be written as

$$\frac{\delta M}{M}\bigg|_V \equiv \frac{\delta M}{\bar{M}_V} = \frac{1}{V}\int_X d^3x\, \delta_m(\boldsymbol{x}) \qquad (2.1.26)$$

Then the dispersion of the mass flucuation is

$$\left\langle \left(\frac{\delta M}{M}\right)^2 \right\rangle_V = \frac{1}{V^2}\int_V d^3x \int_V d^3y\, \langle \delta_m(\boldsymbol{x})\delta_m(\boldsymbol{y})\rangle \qquad (2.1.27)$$

By using the Fourier transformation of the density contrast, we have

$$\langle \delta_m(\boldsymbol{x})\delta_m(\boldsymbol{y})\rangle = (2\pi)^{-3}\int d^3k\, P_m(k)e^{i\boldsymbol{k}\cdot(\boldsymbol{x}-\boldsymbol{y})} \qquad (2.1.28)$$

Inserting this expression

$$\left\langle \left(\frac{\delta M}{M}\right)^2 \right\rangle_V \approx \int^{R^{-1}} \frac{dk\,k^2}{2\pi^2} P_m(k) \qquad (2.1.29)$$

where we have used the following approximation.

$$\frac{1}{V}\int_V d^3x\, e^{i\boldsymbol{k}\cdot\boldsymbol{x}} \approx \begin{cases} 1 & k < R^{-1} \\ 0 & k > R^{-1} \end{cases} \qquad (2.1.30)$$

This is because the exponential oscillate very radidly for $k > R^{-1}$.


## The power spectrum in redshift space

We have to note that the observed distribution of galaxies is not true distribution in the physical space, but is the distribution in the space with distance determined by the redshift. The distance $d_{\mathrm{rss}}$ in this space is defined as.

$$d_{\mathrm{rss}} \equiv \frac{cz}{H_0} + \frac{v_{\mathrm{pec},//}}{H_0} \qquad (2.1.31)$$

where $v_{\mathrm{pec},//}$ is the component of the peculiar velocity in the direction of the line of sight. Since the peculiar velocity is at most of the order of $1000\,\mathrm{kms^{-1}}$, the spatial distribution of galaxies closer than about 14 Mpc will be influenced by this effect (for $H_0 \sim 70\,\mathrm{kms^{-1}Mpc^{-1}}$) and deviate from actual spatial distribution. This also bring about the deviation of the matter power spectrum as follows.

Let denote the position vector in redshift space as $\vec{s}$ and the position vector in actual space as $\vec{x}$. Since the velocity component in the line of sight can be written as $u(r) = \vec{n} \cdot \vec{v}_{\mathrm{pec}} = \mu v_{\mathrm{pec}} \vec{n}$ ($\vec{n}$ is a unit vector in the viewing direction, $\mu$ is the cosine of the line of sight direction and the velocity direction), we have the following relation between real position vectors in real space and redshift space.

$$\vec{s} = \vec{x} + \mu \frac{v(r_{\mathrm{pec}})}{H} \vec{n} \qquad (2.1.32)$$

Since the matter density is conserved under the coordinate transformation, we have

$$1 + \delta_s(\vec{s}) = (1 + \delta(\vec{x}))\, J^{-1} \qquad (2.1.33)$$

where $J = \partial(\vec{s})/\partial(\vec{x})$ is the Jacobian of the transformation. We transform this relation to Fourier space to have

$$\tilde{\delta}_s(\vec{k}) = \int d^3 s\, \delta_s(\vec{s}) = \tilde{\delta}(\vec{x}) + \int d^3 x\, e^{-i\vec{k}\cdot\vec{x}} \left(e^{-i\mu \frac{v_{\mathrm{pec}}}{H} \vec{k}\cdot\vec{n}} - 1\right)(1 + \delta(\vec{x}))$$

Now, we assume that the typical velocity is much smaller than the speed of light and Tayor expand the exponential, leaving only the first-order term, and expand the fluctuation using a plane wave, Using the fact that the velocity field and the wave number vector are parallel.

$$-i\mu \frac{v_{\mathrm{pec}}}{H} \vec{k} \cdot \vec{n} = -i\mu^2 \frac{kv}{H} = \mu^2 f(\Omega_m)\delta \qquad (2.1.34)$$

In the last step, we have used the relation $\dot{\delta} = -ikv$ that is the result of linear density perturbation theory. Also, we used the following notation.

$$f(\Omega_m) = \frac{d\ln \delta}{d\ln a} \qquad (2.1.35)$$

This function is known as the linear growth rate and is known to be well approximated as $f(\Omega_m) \propto \Omega_m^{0.6}$ in general relativity. From the above, it can be seen that within the range of linear approximation, there is the following relationship between density fluctuations in redshift space and density fluctuations in actual space.

$$\delta_s(\vec{s}) = \left(1 + f\mu^2\right)\delta(\vec{x}) \qquad (2.1.36)$$

The relation between power spectrum in real space and redshift space is

$$P_s(k, \mu) = \left(1 + f\mu^2\right)^2 P(k) \qquad (2.1.37)$$

This change in the power spectrum (in the linear case) due to redshift is called the Kaiser term. As can be seen from this expression, this term has the effect of squeezing the distribution in the real space in the direction perpendicular to the line of sight.

On the other hand, on a small scale, the density fluctuation is large and becomes a nonlinear, and the distribution is rather stretched in the viewing direction. This effect is clearly visible in the Nearby Galaxy Survey, and is called the Finger of God effect, which can also be seen in fA2 survey in Figure 2.6.

### 2.1.4. *Dark matter and structure formation*

As mentioned above, what is actually observed as a material density fluctuation is a CMB temperature fluctuation of about $10^{-5}$. Until the universe clears up, CMB photons and baryonic matter interact strongly, so the amplitude of density fluctuations in baryonic matter is also similar. This fluctuation begins to grow when the universe clears up, but it is known that as long as the amplitude of the fluctuation is small, it grows in proportion to the scale factor (will be shown later). Since the redshift of the universe when it clears up is about 1090, it means that it can only grow by 1090 until $z = 0$. Even if the fluctuation that was $O(10^{-5})$ at $z = 1090$ grows about 1000 times, it will only become $O(10^{-2})$, and it will never grow into the galaxy we see today. To form galaxies we need more gravity to accelerate growth. More gravity is achieved with more mass but the mass must not be visible by electromagnetic observations. The dark matter provides the invisible mass.

The existence of dark matter was already suggested in the 1930s from the motions of member galaxies in galaxy clusters. Galaxy clusters are the largest structures in the universe, made up of thousands of galaxies that are bound together by each other's gravity to maintain a certain shape. The shape of a galaxy cluster is maintained by the balance between self-gravity and the pressure caused by the velocity dispersion of member galaxies. Therefore, by measuring the velocity of each galaxy in a galaxy cluster and examining its dispersion, it should be possible to estimate the gravity of the galaxy cluster and the total mass of the galaxy cluster. Based on this consideration, in the 1930s, the Swiss astronomer Zwicky measured the velocity dispersion of member galaxies for a galaxy cluster seen in the direction of the constellation Coma. As a result, he found that the measured velocity dispersion was too large to be explained by mass estimated from the brightness of the member galaxies alone and predicted that a large amount of matter that does not emit light exists within the galaxy cluster. In addition, the existence of dark matter is supported by X-ray observations and gravitational lens observations, and it is extremely difficult to doubt its existence. Various observations have shown that the mass density of dark matter throughout the universe is several times that of baryonic matter. Although dark matter has not yet been directly discovered, structure formation in the universe is discussed asuming the existence of dark matter.

## Structure formation based on Dark matter

Dark matter does not have electromagnetic interactions, so unlike baryonic matter, it can begin to grow at a redshift of about 3400, when matter becomes dominant. By the time the universe clears up, these fluctuations have grown to create slightly dense spherical regions called a dark matter halos, and as the universe clears up, baryonic matter falls into the center of the halo and rapidly increase its density. In this way, it becomes possible to create the galaxies observed to date and their spatial distribution. However, the mass function of dark matter halos (the number of haloes with mass $M$ per unit volume) varies greatly depending on the properties of dark matter, and the way structure formation progresses depends on these differences.

Dark matter is classified into cold dark matter (CDM) and hot dark matter (HDM). This classification depends on whether the velocity dispersion of dark matter particles is non-relativistic (CDM) or relativistic (HDM) at the point when the universe changes from radiation-dominated to matter-dominated (at the time of equality, see below). In addition, warm dark matter (WDM), which has properties intermediate between CDM and HDM, can be considered, but we will not discuss it here.

If HDM is the main component of dark matter, its fluctuations cannot create a halo on a scale smaller than the distance that HDM particles travel between the Big Bang and the equality time due to the relativistic velocity of hot dark matter particles. This scale is called the free streaming scale and is evaluated as follows.

$$L_{\rm FS} \simeq 250 \left( \frac{0.1 {\rm eV}}{m} \right) \ {\rm Mpc} \qquad (2.1.38)$$

where $m$ is the mass of the HDM particle. Therefore, in structure formation based on HDM, structures larger than the free streaming scale are first formed, and then they are successively fragmented to form smaller structures. This way of making the structure is called the top-down scenario.

On the other hand, in the case of CDM, there is no such restriction and abundant small dark matter halos are formed, so small-scale objects are formed first, and then they aggregate and coalesce to form larger-scale structures one after another. This is called a bottom-up scenario.

In this way, HDM and CDM have completely different histories of structure formation.

At the very beginning of the universe, neutrinos were in thermal equilibrium with other particles and they have comparable number density to CMB phtons. Therefore, if neutrinos have masses of about 10 eV, they can be the main component of dark matter, but observations such as KAMIOKANDE show that they only have masses of less than 0.1 eV. and thus they cannot be the main component of dark matter.

### 2.1.5. *Observational evidences of CDM based structure formation*

Simulations of structure formation based on CDM can well explain the large-scale spatial distribution of the observed Universe, and candidates for protogalaxies have already been observed in the Universe with redshifts of 10 or more. These observations support that most of the dark matter is thought to be CDM. Numerical simulations also indicate that the averaged density distribution of the virialized halo created by dark matter can be well approximated by the following equation regardless of its scale.

$$\rho(r) = \frac{\rho_s}{(r/r_s)(1 + r/r_s)^2} \tag{2.1.39}$$

This density distribution is called the Navarro–Frenck–White (NFW) profile, named after the researcher who discovered it through simulation. The measurements of mass distribution of clusters by weak gravitational lensing show that the averaged mass profile can be well approximated by NFW profile.[5]

---

[5]The density distribution of smaler regions such as galaxies and the central region of clusters deviate from NFW profile by the effects of star formation and supernovae explosions.

CDM is assumed to be an elementary particle with a mass of about 100 GeV that has only weak interactions, but it has not been detected in experiments to date. Such elementary particles do not exist in the current standard theory of elementary particles, so if they were detected, they would have a major impact on particle physics. If a black hole with a mass of $10^{-17} \sim 10^{-11} M_\odot$ (referred to as a primordial black hole)[6] was created in the early universe, it would behave as a CDM and the formation of the structure of the universe can be explained without introducing an extra elementary particles.

In addition to the above observations, observations of light elements such as helium and lithium, which originated in the early universe, also play a major role in verifying the Big Bang theory, but this book does not discuss the early universe, so we will not discuss the light element synthesis.

## 2.2. Homogeneous and Isotropic Model of an Expanding Universe

As discussed above large scale galaxy surveys suggest that our universe is homogeneous and isotropic when averaged over about 150 Mpc or larger region. Furthermore, CMB temperature is isotropic with an accuracy $10^{-5}$. These indicate strongly that 3-dimensional space of our universe as a whole can be well described by homogeneous and isotropic geometry.

Mathematically, the fact that the space is homogeneous and isotropic means that the Riemann tensor depends only on a certain constant $K$, and can also be written in terms of metric tensor (or Kronecker's delta). This means that considering the symmetry of the

---

[6]Primitive black holes below this range have emitted gamma rays and evaporated, so they cannot become dark matter. Also, primordial black holes larger than this range are excluded because the gravitational lensing effect caused by them has not been observed.

subscripts, the Riemann tensor can be written in the following form.

$$R_{ijk\ell} = K \left( g_{ik} g_{j\ell} - g_{i\ell} g_{jk} \right) \tag{2.2.1}$$

In fact, if we write 3-dimensional Riemann tensor $R_{ijk\ell}$ as $R_{AB}$ where $A = (ij)$, $B = (k\ell)$, the subscripts $A$ and $B$ takes only three components because of symmetries of Riemann tensor ($R_{ijk\ell} = R_{k\ell ij}$, $R_{ijk\ell} = -R_{jik\ell}$). Furthermore $R_{AB} = R_{BA}$ can be regarded as a symmetric $3 \times 3$ matrix. Thus, it gives a symmetric mapping in 3-dimensional space.

$$X^A \to R^A_B X^B \tag{2.2.2}$$

where $X^A$ is a 3-dimensional vector in 3-dimensional space. Isotropy of the space shows that all eigenvalues of this matrix should be the same. Otherwise there will be a special eigenvalue $\lambda_{(ab)}$ and the corresponding vector $\lambda^c = \epsilon^c_{ab} \lambda^{(ab)}$ which contradicts with the isotropy of the space. Therefore, this matrix is proportional to the unit matrix. Namely,

$$R^A_B \propto \delta^A_B \tag{2.2.3}$$

Come back to the original subscripts, this can be written as

$$R^{ab}{}_{cd} = 2K \delta^{[a}_{c} \delta^{b]}_{d} = K \left( \delta^a_c \delta^b_d - \delta^b_c \delta^a_d \right) \tag{2.2.4}$$

where Bianchi identity shows that the proportional constant $K$ is a constant.

Contracting the Riemann tensor, we have

$$R_{ij} = 2K g_{ij} \tag{2.2.5}$$

Let us calculate this equation explicitly by choosing a convenient coordinate. We can choose any point as a origin by homogeneity and isotropy of space. Then an arbitrary point in space can be specified by the distance $r$ and the directions $(\theta, \phi)$ from the origin. The angular coordinates $\theta$ and $\phi$ are defined by the angles from the arbitrary chosen $z$ axis and $x$ axis, respectively. The radial coordinate $r$ is chosen so that the area of the spherical surface with radius $r$ is $4\pi r^2$.

Points with the same radius $r$ are on the surface of a sphere, so a line element in space can be written as follows.

$$d\ell^2 = e^{2\lambda(r)}dr^2 + r^2\left(d\theta^2 + \sin^2\theta d\phi^2\right) \qquad (2.2.6)$$

where $\Lambda(r)$ is an arbitrary function of the radial coordinate. Ricci tensor can be calculated by this metric tensors as

$$R_{rr} = \frac{2}{r}\frac{d\lambda}{dr}, \quad R_{\theta\theta} = 1 - e^{-2\lambda} + re^{-2f}\frac{d\lambda}{dr}, \quad R_{\phi\phi} = \sin^2\theta R_{\theta\theta} \qquad (2.2.7)$$

Therefore, we obtain the following two equations from (2.2.5).

$$\frac{\lambda'}{r} = 2Kg_{rr} = 2Ke^\lambda \qquad (2.2.8)$$

$$1 + \frac{1}{2}re^{-\lambda}\lambda' - e^{-\lambda} = 2Kg_{\theta\theta} = 2Kr^2 \qquad (2.2.9)$$

These two equations immediately give

$$e^{-\lambda} = 1 - Kr^2 \qquad (2.2.10)$$

Thus, the line element of a homogeneous and isotropic space can be written as follows.

$$d\ell^2 = \frac{dr^2}{1 - Kr^2} + r^2\left(d\theta^2 + \sin^2\theta d\phi^2\right) \qquad (2.2.11)$$

From the definition of $K$, $R = 6K$. Therefore, $K > 0$ is called a closed space, $K = 0$ is a flat space, and $K < 0$ is an open space.

The fact that this space is expanding from moment to moment can be expressed using a function of time $a(t)$ that represents the expansion behavior.

$$d\ell^2(t) = a^2(t)d\ell^2 \qquad (2.2.12)$$

$a(t)$ is called as the scale factor. This means that the spatial coordinate $(x^i)$ does not change even if the space expands. Such coordinates are called a co-moving coordinate system. The line element of a general curved space time that reduces to the form (2.2.11) in a constant $t$ plane can be written as

$$ds^2 = g_{c0}dt^2 + 2g_{0i}dtdx^i + a^2(t)d\ell^2 \tag{2.2.13}$$

Let us take the proper time of galaxies at rest $(dx^i = 0)$ in space as the coordinate time $t$. Then definition of the proper time $d\tau^2 = -ds^2 = g_{00}dt^2$ gives us

$$g_{00} = -1 \tag{2.2.14}$$

We also suppose the motion $dx^i = 0$ is a free fall motion, Then th 4-velocity of a galaxy is always $U^\mu = (1,0,0,0)$ and the spatial components of the geodesic equation become

$$\Gamma^i{}_{00} = 0 \tag{2.2.15}$$

Here,

$$\Gamma^i{}_{00} = \frac{1}{2}g^{ij}g_{j0,0} \tag{2.2.16}$$

Thus, if we choose initially $g_{i0} = 0$, $g_{i0} = 0$ is always satisfied.

In the end the line element of a honomeneous and isotropic expanding universe may be written as follows.

$$ds^2 = -dt^2 + a^2(t)\left[\frac{dr^2}{1 - Kr^2} + r^2d\Omega^2\right] \tag{2.2.17}$$

The spacetime described by this line element is called as Friedmann–Robertson–Walker (FRW) spacetime.

As a radial coordinate, the following coordinate is also used.

$$\chi(r) = \int \frac{dr}{\sqrt{1 - Kr^2}} = \begin{cases} \dfrac{1}{\sqrt{K}}\sin^{-1}\sqrt{K}r & K = +1 \\[2mm] r & K = 0 \\[2mm] \dfrac{1}{\sqrt{-K}}\sinh^{-1}\sqrt{-K}r & K = -1 \end{cases} \tag{2.2.18}$$

Using this coordinate, the line element takes the following form.

$$ds^2 = -dt^2 + a^2 \left[ d\chi^2 + r^2(\chi) \left( d\theta^2 + \sin^2\theta d\phi^2 \right) \right] \qquad (2.2.19)$$

The distance $\chi$ is also called the comoving distance.

### 2.2.1.  *redshift*

Due to the expansion of the universe, electromagnetic waves from distant galaxies are redshifted and their wavelengths are lengthened. As explained later, there are several definitions of distance in cosmology, but the actual observable is redshift. Let's derive the redshift in the FRW universe.

The energy of a photon with 4-momentum $\vec{k}$ measured by an observer $\vec{U}$ is given by

$$E = -\vec{U} \cdot \vec{k} \qquad (2.2.20)$$

Thus, if an observer at affine parameter $\lambda_O$ measured electromagnetic wave emitted by a galaxy at $\lambda_S$, then the reshift observed by the obserber can be written as follows.

$$1 + z = \frac{(\vec{U} \cdot \vec{k})(\lambda_S)}{(\vec{U} \cdot \vec{k})(\lambda_O)} \qquad (2.2.21)$$

For simplicity, we assume that the observer and light source are both at rest with respect to the comoving coordinate. Then the 4-velocity of both have components $(1,0,0,0)$. If we take the observer as the coordinate origin, we can assume that the electromagnetic waves propagate in the radial direction, so using the null condition, we have $\vec{k} = (k^0, k^0/a, 0, 0)$. Therefore, the time component of the geodesic equation is.

$$\frac{dk^0}{d\lambda} + \frac{1}{a^2}\Gamma^0_{\chi\chi} k^\chi k^\chi = 0 \qquad (2.2.22)$$

Using $\Gamma^0_{\chi\chi} = a\dot{a}$ and $k^0 = dt/d\lambda$, this equation becomes

$$\frac{dk^0}{dt} + \frac{\dot{a}}{a} k^0 = 0 \qquad (2.2.23)$$

This shows $k^0 \propto 1/a$, and the ratio of the energy $h\nu_S$ ($h$ is Planck's constant) when emitted and the energy $h\nu_O$ when received is given as

$$\frac{\nu_O}{\nu_S} = \frac{k_O^0}{k_S^0} = \frac{a(\chi_O)}{a(\chi_S)} \tag{2.2.24}$$

Thus, we have the following redshift formula.

$$1 + z = \frac{\lambda_O}{\lambda_S} = \frac{a(\chi_O)}{a(\chi_S)} \tag{2.2.25}$$

The derivation given here is convenient to take Doppler shift into account.

### 2.2.2. *Friedmann equation*

As mentioned above, due to the requirement of homogeneity and isotropy, all metric components except the scale factor can be determined. The constant $K$ that represents the curvature of space is determined by the initial conditions. The equation followed by the scale factor can be found from the Einstein equation.

$$G_{\mu\nu} = 8\pi G T^{\mu\nu} \tag{2.2.26}$$

Using the metric tensor in the Robertsn–Walker spacetime, Enstein tensor can be calculate as follows.

$$G_{00} = 3\left[\left(\frac{\dot{a}}{a}\right)^2 + \frac{K}{a^2}\right] \tag{2.2.27}$$

$$G^i_j = -\left[2\frac{\ddot{a}}{a} + \left(\frac{\dot{a}}{a}\right)^2 + \frac{K}{a^2}\right]\delta^i_j \tag{2.2.28}$$

By the definition of the energy-momentum tensor, the right-hand side of the Einstein equation (2.2.26) is expressed by the energy density

$\rho$ and pressure $P$ as follows.

$$T_{00} = \rho(t), \ T_j^i = P(t)\delta_j^i \tag{2.2.29}$$

Finally, we have the following two equations which given the expansion of the universe.

$$\left(\frac{\dot{a}}{a}\right)^2 + \frac{K}{a^2} = \frac{8\pi G}{3}\rho \tag{2.2.30}$$

$$\frac{\ddot{a}}{a} = -\frac{4\pi G}{3}(\rho + 3P) \tag{2.2.31}$$

Here, we considered a perfect fluid form as the energy momentum tensor. These equations are called Friedmann equations. From the second equation, we can see that the acceleration $\ddot{a}$ is positive only when $\rho + 3P < 0$ is satisfied. Also, from these two equations, the following equations for energy density and pressure can be obtained.

$$\dot{\rho} + 3\frac{\dot{a}}{a}(\rho + P) = 0 \tag{2.2.32}$$

This formula is also derived from the conservation law for the energy momentum tensor. In cosmology, except for the very beginning of the universe, it is sufficient to assume the following simple form as an equation of state.

$$P = \omega\rho \tag{2.2.33}$$

where $\omega$ is called the equation of state parameter and depends on the redshift in general, however most of cases can be treated as a constant.

**Energies in the universe**

There are three types of energy considered in this book.

**1.** $w = 0$, Non-relativistic energy (Dark matter and baryonic matter)

Energy of particles whose velocity is sufficiently small compared to the speed of light and its kinetic energy is negligible compared to its rest mass energy. In this case, the pressure can be ignored, so $w = 0$ is a good approximation. It can be seen from equation (2.2.32) that the energy density is inversely proportional to the third power of the scale factor.

$$\rho_m = \frac{\rho_{m,0}}{a^3} = \rho_{m,0}(1+z)^3 \qquad (2.2.34)$$

Here, we set the current scale factor as 1. We will continue to do so unless otherwise specified. It should also be noted that in the high temperature conditions where thermal energy dominates the rest mass energy, then a non-relativistic particle at low temperatures can become relativistic particle. For example, the mass of an electron is 0.51 MeV, which corresponds to a temperature of about 6 billion degrees. Therefore, in the early universe at temperatures above 6 billion degrees, electrons are regarded as relativistic particles.

**2. $w = 1/3$, relativistic energy**

The energy of a particle moving at the speed of light or a speed sufficiently close to it. The energy of a particle whose rest mass energy is 0 or so small that it can be ignored compared to the thermal energy of the universe. In this case, $w = 1/3$ is derived from statistical mechanics. According to the conservation law, the energy density is inversely proportional to the fourth power of the scale factor.

$$\rho_r = \frac{\rho_{r,0}}{a^4} = \rho_{r,0}(1+z)^4 \qquad (2.2.35)$$

**3. $w \le -1/3$, Dark energy**

The energy that satisfies $\rho + 3P < 0$, that is, $w < -1/3$, is called dark energy. In general, the coefficient $w$ depends on time. For example, the following form is sometimes used with $w_1$, $w_a$ as constants.

$$w(a) = w_0 + (1-a)w_a \qquad (2.2.36)$$

In this case, the conservation law gives

$$\rho_{DE}(z) = \rho_{DE,0}a^{-3(1+w_0+w_1)}e^{-3w_1(1-a)} \qquad (2.2.37)$$

The cosmological constant corresponds to the case $w_0 = -1$, $w_a = 0$, and the energy density is a constant in time.

One goal of observational cosmology is to determine these coefficients. At present, the following limitations have been obtained from CMB observations.

$$w_0 = -0.957 \pm 0.080, \quad w_1 = -0.29^{+0.32}_{-0.26} \qquad (2.2.38)$$

(Planck 2018 results. VI Cosmological parameters). This is not conradict with the case $w_0 = 1$, $w_1 = 0$ that is the cosmological constant. Therefore, we consider only the cosmological constant as dark energy, and the energy density of dark energy is written as $\rho_\Lambda$. If dark energy is a cosmological constant, the current accelerated expansion will continue forever. However, if $w_1 \neq 0$, or $w_a \neq 0$ is confirmed observationally, it suggests the possibility that dark energy is the energy associated with presently unknown elementary particle.

The expansion of the universe is controlled by the dependence of three types of energy densities on scale factors. It can be seen that the main components of the universe are relativistic matter, non-relativistic matter, and dark energy in the order of time. In this book, unless otherwise specified, we will not consider the influence of radiation on the expansion of the universe.

## 2.3. Cosmological Parameters

The parameters that characterize the FRW universe are called cosmological parameters.

First, the Hubble parameter that determines the expansion rate is defined as follows.

$$H(t) \equiv \frac{\dot{a}}{a} \qquad (2.3.1)$$

When indicating the current value, add a subscript 0, such as $H_0$.

Since the Hubble parameter has the dimension of the inverse of time, we can define the following quantity with the dimension of density called the critical density.

$$\rho_{cr,0} = \frac{3H_0^2}{8\pi G} \simeq 1.88 \times 10^{-29} h^2 \, \text{g cm}^{-3}$$
$$\simeq 2.76 \times 10^{11} h^2 \, M_\odot \, \text{Mpc}^{-3} \qquad (2.3.2)$$

Here, $M_\odot \simeq 1.99 \times 10^{33}$ g is the solar mass. Similarly, the density parameter at any scale factor value is defined using the Hubble parameter at any scale factor value. The energy density normalized by the critical density is called the density parameter and is written as follows.

$$\Omega_X(a) = \frac{\rho_X(a)}{\rho_{cr}(a)} \qquad (2.3.3)$$

The current density parameters are written as $\Omega_{X,0} = \Omega_X(a = 1)$. The subscript $X$ is the type of energy density described as follows. We write non-relativistic matter $X = m$, radiation $X = r$, or cosmological constant $X = \Lambda$. Below, we will mainly consider the period long after the universe clears up, so we will ignore the contribution of radiation unless otherwise stated. The acceleration of the expansion is chracterized by the deceleration parameter defined below, but we donot use the parameter in this book.

$$q \equiv -\frac{1}{H^2}\frac{\ddot{a}}{a} \qquad (2.3.4)$$

The following quantity is sometimes used and is called curvature density parameter.

$$\Omega_K(a) = -\frac{K}{a^2 H^2} \qquad (2.3.5)$$

We can express Friedmann equation in terms of these parameters as follows.

$$H^2 = H_0^2 \left( \frac{\Omega_{m,0}}{a^3} + \Omega_{\Lambda,0} - \frac{K}{a^2 H_0^2} \right) \qquad (2.3.6)$$

This equation holds true at any time, but especially considering both sides at the current time $a = 1$, the left side is $H_0^2$. Therefore, the relationship between curvature and energy density is derived.

$$\Omega_{K,0} = -\frac{K}{H_0^2} = 1 - \Omega_{m,0} - \Omega_{\Lambda,0} \tag{2.3.7}$$

Therefore, the sum of density parameters is 1, equivalently the sum of various energy densities is equal to the critical density in the case of flat universe. As mentioned above, this condition is met with an accuracy of better than $10^{-2}$, so from now on, unless otherwise stated, we also ignore the curvature and cosnisder only the case $K = 0$.

### 2.3.1. *Expansion law in the matter and cosmological constant dominated era*

Since we mainly focus our attention to the era after the matter energy density dominates the universe, let us derive the expansion law of the universe in this case.

The condition that the energy density of matter and the energy density of the cosmological constant in the right side of the Friedmann equation gives the redshift when the two quantities are equal

$$1 + z_\Lambda = \frac{1}{a_\Lambda} = \left(\frac{\Omega_{\Lambda,0}}{\Omega_{m,0}}\right)^{1/3} \simeq 1.29 \tag{2.3.8}$$

The red shift $z_\Lambda \simeq 0.29$ corresponds to about 10.3 billion years after the Big Bang.

Neglecting the radiation, the Friedmann equation is written as

$$\dot{a}^2 = H_0^2 \left(\frac{\Omega_{m,0}}{a} + \Omega_{\Lambda,0}a^2\right) \tag{2.3.9}$$

Thus, we have

$$H_0 t = \int_0^a \frac{\sqrt{a}\,da}{\sqrt{\Omega_{m,0} + \Omega_{\Lambda.0}a^3}} \tag{2.3.10}$$

The integral on the right hand side of this equation is calculated using $\left(\sinh^{-1} x\right)' = \frac{1}{\sqrt{x^2+1}}$ to give

$$a(t) = \left(\frac{\Omega_{m,0}}{1-\Omega_{m,0}}\right)^{-1/3} \sinh^{2/3}\left(\frac{3}{2}\sqrt{1-\Omega_{m,0}}H_0 t\right) \qquad (2.3.11)$$

From the above expression, we can easily obtain the expansion law in the period where the cosmological constant can be ignored,

$$a(t) = \Omega_{m,0}^{1/3}\left(\frac{3}{2}H_0 t\right)^{2/3} \qquad (2.3.12)$$

On the other hand, in the limit where the matter can be neglected, we have the following expansion law.

$$a(t) = a_\Lambda \exp\left(H_0\Omega_{\Lambda,0}^{1/2}(t - t_\Lambda)\right) \qquad (2.3.13)$$

We can express the time as a function of the scale factor as

$$t(a) = \frac{2}{3H_0\sqrt{1-\Omega_{m,0}}} \ln\left[\left(\frac{a}{a_\Lambda}\right)^{3/2} + \sqrt{1 + \left(\frac{a}{a_\Lambda}\right)^3}\right] \qquad (2.3.14)$$

If we set $a = 1$ in the above formula, we can get the current age of the universe.

$$t_0 = \frac{2}{3H_0\sqrt{1-\Omega_{m,0}}} \ln\left[\frac{1 + \sqrt{1-\Omega_{m,0}}}{\sqrt{\Omega_{m,0}}}\right] \simeq 13.8\,\text{Gyr} \qquad (2.3.15)$$

## 2.3.2. *Growth of density flactuation in an expanding universe*

In the previous section, we looked at the distribution of galaxies currently observed in the universe. This distribution is the result of gravitational growth of density flctuations in the early universe. Let us now consider the structure formation in the expanding universe in some detail.

The following three elements play important roles in structure formation:

- Initial spectrum of density perturbation
- Gravity
- Expansion of the universe

The shape of the initial spectrum tells us at what scale and how large density fluctuations are prepared. Gravity promotes the growth of density fluctuations, and cosmic expansion inhibits growth. Dark matter is the main source of gravity and dark energy plays an important role in the expansion.

First, we consider the initial power spectrum,

## Harrison–Zeldovich Spectrum

Although we do not discussed in this book, it is believed that a rapid accelerated expansion occurred at the very beginning of the universe. This expansion is called inflationary expansion or simply inflation and quantum fluctuations in the energy density during the inflation is thought to be the origin of fluctuation which bring about the present structures. According to this scenario, the following power spectrum is predicted as the initial power spectrum and is known to explains the observed matter power spectrum well at the current universe.

$$P_m(k) \propto k^{n_s}, \quad n_s \simeq 1 \tag{2.3.16}$$

In particular, the case $n_s = 1$ is called the Harrison–Zeldovich (HZ) spectrum which is proposed by Harrison and Zeldovich as the initial power spectrum before the inflationary scenario was proposed. The reason why HZ spectrum is special is as follows.

From (2.3.16) and noticing $k \propto R^{-1} \propto M^{-1/3}$, we can estimate the dispersion of the mass fluctuation as a function of mass as follows.

$$\left\langle \left(\frac{\delta M}{M}\right)^2 \right\rangle_V \approx \int^{R^{-1}} \frac{4\pi k^2}{2\pi^2} P_m(k) \propto k^{3+n_s} \propto M^{-(3+n_s)/3}$$

(2.3.17)

Let us consider the fluctuations of the Newtonian potential $\delta\Phi$ induced by mass fluctuation $\delta M$.

$$\delta\Phi = -\frac{G\delta M}{R}$$

(2.3.18)

Then the dispersion of potential fluctuation can be written as

$$\left\langle (\delta\Phi)^2 \right\rangle \sim \left\langle \left(\frac{G\delta M}{R}\right)^2 \right\rangle \propto \left\langle \left(\frac{G\delta M}{M^{1/3}}\right)^2 \right\rangle \propto M^{4/3} \left\langle \left(\frac{G\delta M}{M}\right)^2 \right\rangle$$

$$\propto M^{-\frac{n_s-1}{3}}$$

(2.3.19)

Therefore, when $P_m \propto k$, the wavenumber dependence of the dispersion of the gravitational potential by mass fictuation disappears, and the dispersion of the gravitational potential due to fluctuations is constant regardless of the scale of fluctuations. Such a spectrum is called a scale-free spectrum.

This spectrum can also be predicted from the following simple considerations. If $P_m \propto k^{n_s}$ and $n$ is much larger than 1, the potential fluctuation is very large at large wavenumbers(small scales), creating a universe full of black holes in small scales. On the other hand, if $n$ were much smaller than 1, the fluctuations would be very large on a large scale with small wavenumbers, resulting in a globally inhomogeneous universe. In this way, a power spectrum close to the Harrison–Zeldovich spectrum is needed to explain the current universe. Inflation theory naturally provides such a spectrum and is regarded as the standard theory of the early universe. Inflation theory predicts not only fluctuations in the matter density but also the existence of gravitational waves with a characteristic spectrum. One of the goals of observational cosmology is to approach the beginning of the universe through observation, and the inflation theory can be verified through precise observations of matter power spectrum and primordial gravitational waves in future.

### 2.3.3.  *Gravitational growth of density perturbation*

Given the initial matter spectrum, the fluctuations evolve under the influences of self-gravity and the expansion of the universe, and the evolution equation takes the following simple form in the linearized theory.

$$\ddot{\delta} + 2\frac{\dot{a}}{a}\dot{\delta} - 4\pi G \rho_b \delta = 0 \qquad (2.3.20)$$

We note here that $\rho_b$ is the background mean density that govens the expansion of the universe, and thus it differs depending on the era we are considering.

The second term in the left-hand-side of this equation expresses the resistance by the expansion of the universe. Thus, if there is no expansion, then the density contrast grows exponentially. By solving this equation with the knowledge of the time dependence of the scale factor, the growth of fluctuations can be given as a function of time or as a function of the scale factor. For example, the universe expands as $a(t) \propto t^{2/3}$ in the matter dominant era, so this equation can be easily solved as

$$\delta \propto t^{2/3} \propto a(t), \quad \delta \propto t^{-1} \qquad (2.3.21)$$

Thus, the density fluctuations grow in the same way as the scale factor in this case. A solution that increases in time is called a growing mode, and a solution that decays is called a decaying mode. The time-dependent part of the growing mode of density fluctuation is written as $D_+$. We write the solution in the following form.

$$\delta(t, \vec{x}) = D_+(t)v(\boldsymbol{x}) \qquad (2.3.22)$$

where a time-independent function $v(\boldsymbol{x})$ is choosen to satisfie the initial spectrum. The following approximate expression is known for the growth mode function in the general case with a cosmological constant after the material-dominated period (Lahav *et al.*, 1991).

$$D_+(a) \simeq \frac{5}{2} a \Omega_m(a) \left[ \Omega_m^{4/7}(a) - \Omega_\Lambda(a) \right.$$

$$\left. + \left(1 + \frac{\Omega_{m,}(a)}{2}\right) \left(1 + \frac{\Omega_\Lambda(a)}{70}\right) \right]^{-1} \tag{2.3.23}$$

The following expression called growth rate is also often used for growing mode.

$$f(z) = \frac{d \ln D_+(a)}{d \ln a} \tag{2.3.24}$$

Approximate formulas for this are also known,

$$f(a) \simeq \Omega_m^{4/7}(a) + \frac{\Omega_\Lambda(a)}{70} \left[ 1 + \frac{\Omega_m(a)}{2} \right] \tag{2.3.25}$$

In the case with $\Omega_n + \Omega_\Lambda = 1$, the exact solution is known, but the above approximate solution is often sufficient.

## DM density fluctuation around radiation-matter equality

When we consider the universe near the equality time between radiation and matter, we can neglect the curvature term and the cosmological constant but cannot ignore the contribution by radiation density in the Friedmann equation. Therefore, the relevant Friedmann equation can be written as

$$\left(\frac{\dot{a}}{a}\right)^2 = \frac{8\pi G}{3} (\bar{\rho}_m + \rho_r) \tag{2.3.26}$$

$$\frac{\ddot{a}}{a} = -\frac{4\pi G}{3} (\bar{\rho}_m + 2\rho_r) \tag{2.3.27}$$

where $\rho_r$ is the radiation density, $\bar{\rho}_m$ is the averaged matter density. The matter fluctuation is written as

$$\rho_m(t) = \bar{\rho}_m(t)\left(1 + \delta_{DM}\right) \tag{2.3.28}$$

Since baryonic matter do not grow until the decoupling $z \simeq 1090$, and we are here interested in the period before the decoupling, we consider only the fluctuation of dark matter. Then the evolution equation of DM fluctuation takes the following form.

$$\ddot{\delta}_{DM} + 2\frac{\dot{a}}{a}\dot{\delta}_{DM} - 4\pi G\bar{\rho}_m\delta_{DM} = 0 \tag{2.3.29}$$

It is convenient to rewrite this equation using the matter density parameter $\Omega_m(z) = \frac{8\pi G\bar{\rho}_m(z)}{3H^2(z)}$ as follows.

$$\ddot{\delta}_{DM} + 2\frac{\dot{a}}{a}\dot{\delta}_{DM} - \frac{3}{2}H^2\Omega_m(z)\delta_{DM} = 0 \tag{2.3.30}$$

We consider two limiting case for simplicity.

- The radiation dominated era with $z \gg z_{eq} \sim 3230$
  In this period can ignore the contribution of matter, thus we approximately take $\Omega_m = 0$ and $a(t) \propto t^{1/2}$ in (2.3.30) to find

$$\ddot{\delta}_{DM} + \frac{1}{t}\dot{\delta}_{DM} = 0 \tag{2.3.31}$$

The solution is easily found to be

$$\delta_{DM} = A_1 + A_2\log t \tag{2.3.32}$$

where $A_1, A_2$ are constans. Thus, the dark matter grows only logarithmically. This slow growth of the DM perturbation is called stagspansion.
- The matter dominatec era with $z_{eq} \ll z \ll z_\Lambda$
  In this period, we can safely take $\Omega_m = 1$ and $a \propto t^{2/3}$, then we recover the previous result (2.3.21).

$$\delta_{DM} = B_1 t^{-1} + B_2 t^{2/3} \tag{2.3.33}$$

where $B_1, B_2$ are constants.

### 2.3.4. *Evolution of the primordial spectrum*

Since we now know the evolution of matter density fluctuation, we are ready to understand the evolution of the matter power spectrum. Horizon is the region within which there are causal interactions and thus the boundary of the horizon extends with a speed of light. Thus the horizon distance at a time $t$ is the distance that light travel from Big Bang to the time $t$. The evolution equation (2.3.20) applies only for the fluctuation whose typical scale (one wavelength) is smaller than the horizon distance., and we already know that DM fluctuation inside horizon does not grow in time during the radiation. On the other hand, DM flctuation whose typical scale is larger than horizon distance grow as $a^2$. This can be seen as follows. Consider a flat universe with no cosmological constant (we can safely neglect the effect of the cosmological constant in the period we are now interested in), and a spherical region whose density is uniformly higher than the average density $\rho_b$ by $\rho_b \delta$. Then, the equation for the expansion of the universe in each region can be written as follows:

$$\left(\frac{\dot{a}}{a}\right)^2 = \frac{8\pi G}{3}\rho_b$$

$$\left(\frac{\dot{a}}{a}\right)^2 = \frac{8\pi G}{3}\rho_b(1+\delta) - \frac{K}{a^2}$$

Since the interior of the sphere is dense($\delta > 0$), a positive curvature ($K > 0$) has to be included in order to match the two region smoothly, namely Hubble parameters coincide at the boundary. Subtracting these two expressions gives us

$$\delta = \frac{3K}{8\pi G}\frac{1}{\rho_b a^2} \tag{2.3.34}$$

In the radiation-dominant (RD) period, $\rho_b \propto a^{-4}$ leads to $\delta \propto a^2$ ($\delta \propto a$ in the material-dominated MD period).

In the RD and MD eras, the horizon distance expands in proportion to time, but the scale of fluctuations expands as the universe expands, that is proportional to the scale factor. Since the expansion in these eras is decelerating, $a(t) \propto t^n; n < 1$, typical scales of any density fluctuation become shorter than the horizon distance. The moment when the typical scale (wavelength) of the density fluctuation is equal to the horizon distance is called horizon crossing. We say that the fluctuations is outside the horizon before the horizon crossing and inside the horizon after the horizon crossing. We also say that the fluctuation enters the horizon at the moment of the horizon crossing. Given the comoving wavenumber, one can calculate the redshift of the horizon crossing as follows. The horizon distance can be calculated as the inverse of the Hubble parameter.

$$L_H(z) = H_0^{-1} \left[ \Omega_{m,0}(1+z)^3 + \Omega_{r,0}(1+z)^4 \right]^{-1/2}$$

$$\simeq \begin{cases} H_0^{-1}\Omega_{r0}^{-1/2}(1+z)^{-2} & z \gg z_{eq} \approx 3230 \\ H_0^{-1}\Omega_{m0}^{-1/2}(1+z)^{-3/2} & z_{eq} \ll z \ll z_\Lambda \approx 0.29 \end{cases} \qquad (2.3.35)$$

The condition of the horizon crossing for the flctuation with the wavelength $\lambda_0 = 2\pi/k_0$ is defined as follows.

$$\lambda_0 a_{\text{enter}} = L_H(z_{\text{enter}}) \qquad (2.3.36)$$

Using (2.3.35), we have for $z \gg z_{eq}$

$$\lambda_0 = (1+z_{\text{enter}})L_H(z_{\text{enter}}) = H_0^{-1}\Omega_{r0}^{-1/2}(1+z_{\text{enter}})^{-1} \qquad (2.3.37)$$

or

$$z_{\text{enter}} \approx H_0^{-1}\Omega_{r0}^{-1/2}\lambda_0^{-1} \approx 4.55 \left( \frac{\lambda_0}{1\,\text{Mpc}} \right)^{-1} \qquad (2.3.38)$$

Similary, we can derive the redshift of the horizon crossing for $z_{eq} \ll z \ll z_\Lambda$

$$z_{\text{enter}} \approx 3000 \left( \frac{\lambda_0}{150\,\text{Mpac}} \right)^{-1} \qquad (2.3.39)$$

Therefore, the density fluctuations that formed the observed structure at present universe entered the horizon during the radiation-dominant period and we consider such fluctuations only. Fluctuations entered the horizon during the matter dominant era do not change the growing behavior before and after the horizon crossing. Therefore, the shape of the power spectrum does not change for fluctuations larger than the wavenumber $k_{eq} = a_{eq}H(a_{eq})$ that corrsponds to the scale of the radiation matter equality time.

$$k_{eq} = a_{eq}H_0\sqrt{\frac{\Omega_{m,0}}{a^3} + \frac{\Omega_{r,0}}{a^4}} = H_0\sqrt{\frac{2\Omega_{m,0}^2}{\Omega_{r,0}}} \simeq 0.72\,\Omega_{m,0}h^2\,\mathrm{Mpc}^{-1}$$

(2.3.40)

where we used $a_{eq} = \Omega_{r,0}/\Omega_{m,0}$. When we substitute the observed value $\Omega_{m,0} \simeq 0.12$ in the above expression, we have $k_{eq} = 0.02\,\mathrm{Mpc}^{-1}$ and this corresponds to .the position where the slope of the graph changes in Fig. 2.7.

Now, let us look at shorter fluctuations that entered the horizon in radiation dominant era. Note that the shorter the scale of fluctuation, the faster it crosses the horizon. Once a fluctuation enters the horizon, it cannot grow in time until the radiation-matter eqlity time, so the shorter the scale of the fluctuation, the longer the growth

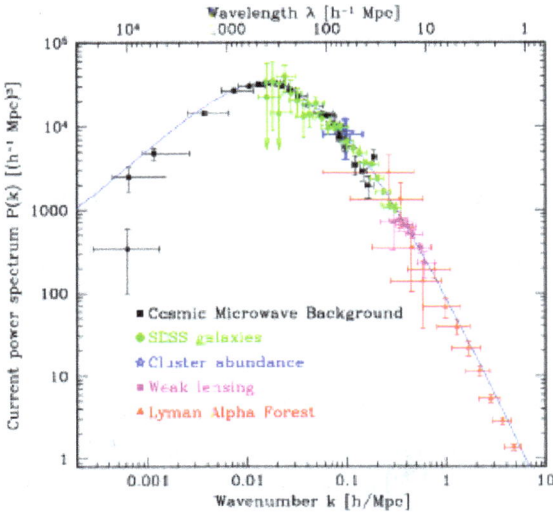

Fig. 2.7. The observed power spectrum at the present universe by various methods.

will be inhibited. Remember the fictuations outside the horizon grows as $a^2$. Consider the fluctuation $\delta_k$ with a fixed wavenumber $k$. Then the horizon distance that the fluctuation enter is $L_H^{-1}(k) \propto k$ (this simply tells you that the higher the wavenumber, the earlier it enters the horizon). Since the scale factor when the fluctuation enters the horizon is given by $a_{\text{enter}}(k) \propto L_H(k)$ and the fluctuation grow until the entry of the horizon

$$\delta_k \propto a_{\text{enter}}^2(k)\delta^{(\text{ini})}(k) \equiv T(k)\delta^{(\text{ini})}(k) \qquad (2.3.41)$$

where $\delta^{(\text{ini})}$ is the initial density contrast and $T(k)$ is called the transfer function. Thus, we find

$$P(k) \propto T(k)^2|\delta^{(\text{ini})}(k)|^2 \propto \begin{cases} k & k \ll k_{\text{eq}} \\ k^{-3} & k \gg k_{\text{eq}} \end{cases} \qquad (2.3.42)$$

The shape of the primordial spectrum is maintained for fluctuations of $k < k_{eq}$, so predictions of inflation can be verified from observations of large-scale fluctuations. There are various influences on the fluctuations of $k > k_{eq}$, such as the BAO mentioned above, dark matter, and the nonlinear growth of gravity, but the general feature of the observed shape of the matter power spectrum in the present universe can be explained in this way.

### 2.3.5. *Structure formation scenario based on CDM*

Since all the discussions so far have been based on linear theory, there has been no information on the amplitude of fluctuations. Information on the amplitude of fluctuations can be obtained from the observations of temperature fluctuations in the microwave cosmic background radiation (CMB). CMB temperature fluctuation is a snapshot of the universe at $z \approx 1090$ (about 380,000 years after the Big Bang). Until the last scattering, CMB photons and baryonic matter interact strongly by Thomson scattering, so the amplitude of density fluctuations in baryonic matter is of the order of $O(10^{-5})$. According to the theory of CMB temperature fluctuation, the temperature fluctuation at the last scattering is related with the potential $\Psi$ by the density fluctuation as follows.

$$\left.\frac{\delta T}{T}\right|_{z=1090} = \frac{1}{3}\Psi \tag{2.3.43}$$

where the potential satisfies Poisson equation.

$$\Delta \Phi = 4\pi G \rho_b \delta \tag{2.3.44}$$

Thus, the matter power spectrum is given by the power spectrum of potential as follows.

$$P_\Phi(k) \propto k^{-4} P_{\delta_m}(k) \tag{2.3.45}$$

It can be shown that there is a relation between CMB power spectrum $C_\ell$ and the potential power spectrum.

$$\frac{\ell^2 C_\ell}{2\pi} = \frac{1}{9}\frac{k^3 P_\Phi(k)}{2\pi^2} \tag{2.3.46}$$

the right-hand side of the equation is usually written as

$$\frac{1}{9}A_s \left(\frac{k}{k_0}\right)^{n_s-1} \tag{2.3.47}$$

where $k_0$ is the reference wave number, and is usually taken as $k_0 = 0.002\,\mathrm{Gpc}^{-1}$. $A_s$ plays the role of the amplitude of the fluctuation. In the case of Harrison–Zeldovich spectrum $n_s = 1$, we have

$$A_s = \frac{9}{2\pi}\ell^2 C_\ell \tag{2.3.48}$$

Planck satelite gives the following values for the amplitude $A_s$ and $n_s$.

$$\log\left(10^{10} A_s\right) \approx 3, \quad n_s \approx 0.968 \tag{2.3.49}$$

We have already seen that the dark matter is needed to form galaxies we see today. Furthermore various observations support that the main component of DM is cold dark matter(CDM). Accepting the existence of CDM, we have the following scenario of the structure formation. Dark matter does not have electromagnetic interactions, so unlike baryonic matter, it can start growing at a redshift of about 3230, when matter becomes dominant. By the time of the last scattering, CDM fluctuations have grown to form a slightly dense spherical region called a dark matter halo from smaller size to large size according to the shape of the power spectrum we have just described. As soon as the universe clears up, baryonic matter falls into a halo and rapidly increases in density at its center.

The early evolution of the bryonic matter after decoupling can be described by the following equation.

$$\ddot{\delta_b} + \frac{4}{3t}\dot{\delta_b} - \frac{2}{3t^2}\delta_{DM} = 0 \qquad (2.3.50)$$

where we used that $a \propto t^{2/3}$. The third term of this equation is determined by the dark matter density contarst $\delta_{DM}$ because baryons are in the gravitational potential generated by the dark matter fluctuation. By subtracting this equation from the dark matter evolution equation (2.3.29) and using $\delta_{DM} = Ca$ for a constant $C$, we find the following equation for the difference $D = \delta_b - \delta_{DM}$.

$$a^2\frac{d^2D}{da^2} + \frac{3}{2}a\frac{dD}{da} = 0 \qquad (2.3.51)$$

This has solution as

$$D(a) = D_1 + D_2 a^{-1/2} \qquad (2.3.52)$$

where $D_1, D_2$ are constantrs. Neglecting the decaying mode and using the initial condition as $\delta_b = 0$ at $a = a_{LS} = 1/1090$, we find

$$\delta_b(a) = \delta_{DM}(a)\left(1 - \frac{a_{LS}}{a}\right) \qquad (2.3.53)$$

After falling into the dark matter halo, baryonic matter eventually forms stars. In this process the cooling by hydrogen molecules plays an important role. The detail of the star formation in CDM halo is complicated and is beyond the scope of this book and reader may consult appropriate textbooks.

## 2.4. Distance-redshift Relation

The proper distance measured at both ends at the same time is not practical in cosmological situation because the galaxies we observe are hundreds of millions or even billions of light years away. Therefore, it is necessary to define distance directly associated with observable. The most useful such distances are the luminosity distance and the angular diameter distance.

### 2.4.1. *Luminosity distance*

Luminosity distance is defined by the observed magnitude of the source. This is only possible when we know that the source is the standard candle. Suppose that the absolute luminosity (energy emitted per unit time) of the galaxy is $L$, this energy is radiated in all directions, and we observe a portion of it. In 3-dimensional Euclidean space, if there is a light source at a distance $D$, the energy flux $f$ that is the energy received per unit time and unit area is expressed as follows.

$$f = \frac{L}{4\pi D^2} \tag{2.4.1}$$

Using this relation we can define a distance called the luminosity distance $D_L$. In other words, when we observe the energy flux $f$ from a distant galaxy whose absolute luminosity is $L$, then we define the luminosity distance to the galaxy as follows

$$D_L = \left( \frac{L}{4\pi f} \right)^{1/2} \tag{2.4.2}$$

Now, let us consider that galaxy $G_1$ at redshift $z_1$ receives the energy radiated by galaxy $G_2$ at redshift $z_2$. Let the comoving radial distances corresponding to each redshift be $\chi_2$ and $\chi_1$, respectively. From isotropy of space we can assume that light travels in the radial direction, the relationship between the time interval and the radial interval on the world line of light is given by the line element (2.2.19) with $d\theta = d\phi = 0$

$$0 = ds^2 = -dt^2 + a^2(t)d\chi^2 \tag{2.4.3}$$

Thus, we have the following relationship between the radial coordinates $\chi$ and redshift.

$$\int_{\chi_1}^{\chi_2} d\chi = \chi_2 - \chi_1 = \frac{1}{H_0} \int_{z_1}^{z_2} \frac{dz}{\sqrt{\Omega_{m,0}(1+z)^3 + \Omega_{\Lambda,0} - k_0(1+z)^2}} \tag{2.4.4}$$

where we used the following relation to replace the integration variable from time to scal factor.

$$\frac{dt}{a} = \frac{da}{a^2 H(a)} = -\frac{dz}{H(z)} \tag{2.4.5}$$

The radial distance between two galaxies $r(z_1, z_2)$ can be calculated by applying the above equation to $\chi_2 - \chi_1$ on the right side of the equation below.

$$r(z_1, z_2) = \begin{cases} \sin(\chi_2 - \chi_1) & K = +1 \\ \chi_2 - \chi_1 & K = 0 \\ \sinh(\chi_2 - \chi_1) & K = -1 \end{cases} \tag{2.4.6}$$

For example, in the case of flat universe, we have

$$r(z_1, z_2) = \frac{1}{H_0} \int_{z_1}^{z_2} \frac{dz}{\sqrt{\Omega_{m,0}(1+z)^3 + (1 - \Omega_{m,0})}} \tag{2.4.7}$$

Now, if the energy radiated by galaxy $G_2$ in the time interval $\delta t_2$ is $\delta E_2$, the luminosity can be written as $L_2 = \delta E_2/\delta t_2$.

From the definition of radial distance this energy spreads over sphere of area $4\pi\left(a(z_1)r(z_1, z_2)\right)^2$ by the time it reaches a galaxy $G_1$ at $\chi = \chi_1$. Thus, the energy flux received by galaxy $G_1$ can be written as follows.

$$f_1 = \frac{\delta E_1/\delta t_1}{4\pi \left(a(z_1)r(\chi_1,\chi_2)\right)^2} \tag{2.4.8}$$

where $\delta E_1$, $\delta t_1$ are the energy and time interval received by galaxy $G_1$. These are related with the original $E_2$ and $\delta t_2$ as follows.

$$\frac{\delta E_2}{\delta E_1} = \frac{\delta t_1}{\delta t_2} = \frac{a(z_1)}{a(z_2)} = \frac{1+z_2}{1+z_1} \tag{2.4.9}$$

These are simply the redshift by the expansion of the universe.

$$f_1 = \left(\frac{1+z_1}{1+z_2}\right)^2 \frac{\delta E_2}{4\pi \left(a(z_1)r(\chi)\right)^2 \delta t_2} \tag{2.4.10}$$

Finally comparing this expression with the definition of the luminosity distance we have

$$D_A(z_1, z_2) = \left(\frac{L_2}{4\pi f_1}\right)^{1/2} = \frac{1+z_2}{(1+z_1)^2} r(\chi_1,\chi_2) \tag{2.4.11}$$

where $r(\chi)$ is given by (2.4.6).

For example, in the case where $K = 0$, $\Lambda = 0$ (Einstein-de Sitter universe), we

$$D_L(z_1, z_2) = \frac{2}{H_0}\frac{1+z_2}{1+z_1}\left[\frac{1}{\sqrt{1+z_1}} - \frac{1}{\sqrt{1+z_2}}\right] \tag{2.4.12}$$

For flat universe with the cosmological constant, the luminosity dsitance is given by

$$D_L(z) = D_L(0, z) = \frac{1}{H_0}(1+z) \int_0^z \frac{dz}{\sqrt{\Omega_{m,0}(1+z)^3 + 1 - \Omega_{m,0}}} \tag{2.4.13}$$

Applying this formula into the distance modulus $m - M$, the acceleration of the cosmic expansion was discovered.

### 2.4.2. *Angular diameter distance*

In Euclidean space, if you know the length of an object, you can measure the distance to it based on how small it appears. The angular diameter distance is defined in this way in the cosmological situation. Therefore, we need the standard ruler that is the object whose absolute size is known.

Let us consider a rod with the proper length $L$ in the direction perpendicular to the line of sight at redshift $z_2$ and radial coordinate $\chi_2$. If the angle at which the rod is viewed from redshift $z_1$ is $\delta\theta$, then the angular diameter distance from redshift $z_1$ to $z_2$ is defined as follows.

$$D_A(z_1, z_2) \equiv \frac{L}{\delta\theta} \qquad (2.4.14)$$

Since $dt_2 = d\chi_2 = d\phi_2 = 0$ along the rod, the proper distance is given by,

$$L = a(z_2)r(z_1, z_2)\delta\theta \qquad (2.4.15)$$

where we take the origin of the spatial coordinate at the observer. Thus,

$$D_A(z_1, z_2) = a(z_2)r(z_1, z_2) = \frac{r(z_1, z_2)}{1 + z_2} \qquad (2.4.16)$$

Therefore, the angular diameter distance and the luminosity distance are related each other as

$$D_A(z_1, z_2) = \left(\frac{1 + z_1}{1 + z_2}\right)^2 D_L(z_1, z_2) \qquad (2.4.17)$$

Figure 2.8 shows the dependence of the luminosity distance (a) and angular diameter distance (b) on cosmological parameters. The angular distance reaches its maximum value at a certain redshift and gradually decreases beyond that point. This means that celestial bodies farther away than a certain redshift appear

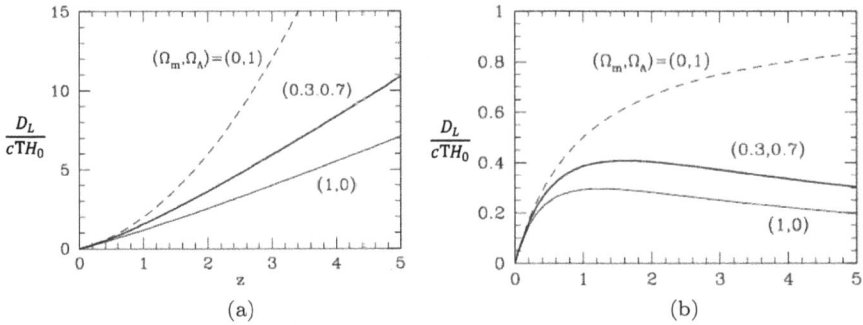

Fig. 2.8. Dependence of luminosity and angular diameter distances on the cosmological parameters.

larger the further away you go. At first glance, this may seem like a contradiction, but, it is not a contradiction because it actually appears that way. This phenomenon occurs because light emitted from an object farther away than a certain redshift is bent by gravity of matter. In fact, when you decrease the matter density parameter, the position of the peak goes a higher redshift and in case there is no matter (De Sitter universe $(\Omega_{m0}, \Omega_{\Lambda,0}) = (0,1)$), the angular diameter distance increases monotonically. If we can directly measure angular or luminous distances at various redshifts, we can determine cosmological parameters. This is exactly the case with the luminosity/redshift relationship of type Ia supernovae.

The angular distance is used to derive cosmological information from gravitational lensing and CMB temperature fluctuations.

## 2.5. Observables in an Inhomogeneous Universe

So far we have considered an homogeneous and isotropic universe as a useful approximation for horizon scale averaged universe. However, lights coming from distant galaxies do not feel the statistical distribution of matter, but propagate on inhomogeneous distribution of local material distribution. This changes the brightness and shape of distant galaxies, In extreme cases, multiple images are observed.

Here, we study general behavior on the propagation of light rays on a realistic inhomogeneous universe.

Inhomogeneous matter distribution creates inhomogeneities in gravitational potential. Electromagnetic waves propagate therein according to geodesic equations. Therefore, in order to investigate the details, it is first necessary to know the cosmological metrics with inhomogeneous matter distribution. Such geometry seems to be very complicated, but the following consideration makes the situation much simpler.

In a cosmological circumstances we are interested in this book, the gravitational field created by an individual celestial body is not very strong except in special circumstances, and the time scale on which the gravitational field fluctuates is sufficiently long compared to the time it takes for electromagnetic waves to propagate in the vicinity of that celestial body. Therefore, it is sufficient to consider cases where the gravitational field is weak and its temporal fluctuations can be ignored. In this case, the line elements in the universe can be treated as small perturbations from a homogeneous and isotropic space-time. In the case of asymptotically flat spacetime as we saw in Chapter 1, the space-time line element in the region far from the gravity source was expressed as follows.

$$ds^2 = -(1 + 2\Psi)\, dt^2 + (1 - 2\Psi)\left(dx^2 + dy^2 + dz^2\right) \qquad (2.5.1)$$

In our cosmological situation, we generalize the line element by taking the expansion of the universe into account as follows.

$$ds^2 = -(1 + 2\Psi)\, dt^2 + a^2(1 - 2\Psi)\left(dx^2 + dy^2 + dz^2\right) \qquad (2.5.2)$$

Here, we restricted ourself to the period long after the last scattering and flat background universe, namely we consider the small perturbation on the flat background. We write the matter distribution ass follows.

$$\rho_b(z) = \rho_{cr}(z)\left(1 + \delta_m(z)\right) \qquad (2.5.3)$$

where the background density $\rho_b = \bar{\rho}_m + \rho_\Lambda = \rho_{cr}$, and we consider the fluctuation of matter distribution $\delta_m$ only and its average vanishes

$$\langle \delta_m \rangle = 0 \qquad (2.5.4)$$

Ideally the averaging is ensamble average, but in practice is the spatial average over a sufficiently large volume, say of the order of 150 Mpc or so. Then by averaging Einstein equation, we can show that the scal factor obey the Friedmann equations within the approximation neglecting the second order of the density fluctuation and the potential generated by the matter fluctuation obeys the following Poisson like equation.

$$\Delta \Psi = 4\pi G a^2 \rho_{cr} \delta \qquad (2.5.5)$$

Note that the above line element is valid even when the density fluctuations are very large. The line element of the above form was found by approximation that the deviation from the flat line element of the universe is sufficiently small, so the Newtonian potential must be sufficiently small. Now, let $\epsilon$ be an infinitesimal quantity and let Newton's potential be on the order of $\epsilon^2$ (In a gravitationally bund system, $\epsilon$ is the typical order of velocity that balances gravity).

$$\Psi = O\left(\epsilon^2\right) \qquad (2.5.6)$$

We can estimate the order of the critical density as follows.

$$G\rho_{cr} = \frac{3H^2}{8\pi} = O\left(\frac{1}{L_H^2}\right) \qquad (2.5.7)$$

Then Poiasson like equation (2.5.5) gives the following order estimation for the density contrast.

$$\delta = O\left(\frac{\epsilon^2}{\ell^2/L_H^2}\right) \qquad (2.5.8)$$

where $\ell$ is the typical scale at which the matter distibution changes. Thus, as far as $\frac{\ell}{L_H} \ll \epsilon \ll 1$ is satisfied, the density contrast could be larger than 1. The linear approximation for the metric is a good approximation even if the matter fluctuation becomes nonlinear as far as the scale of the fluctuations are much smaller than the horizon. For example, the typical scale of galaxy is of the order of $\ell \simeq 10\,\text{kpc}$

and $\ell/L_H \simeq 10^{-8}$ or so. On the other hand, the velocity dispersion of an elliptical galaxies or rotation velocity of a disk galaxy is $\epsilon \simeq 10^{-3}$. Thus, the above line element can be safely applied for galactic scales.

### 2.5.1. *Geodesic equation in an inhomogeneous universe*

In order to study, the light propagation in an inhomoegeous universe, it is convenient to use the following conformally related line element $d\tilde{s}^2$ with the original line element $ds^2$ given by (2.5.2).

$$ds^2 = a^2(\eta)d\tilde{s}^2 \tag{2.5.9}$$

where

$$d\tilde{s}^2 \equiv \tilde{g}_{\mu\nu}dx^\mu dx^\nu$$
$$= -\left(1 + 2\Psi\right)d\eta^2 + \left(1 - 2\Phi\right)\left(dx^2 + dy^2 + dz^2\right) \tag{2.5.10}$$

We have used the conformal time $\eta = \int^t dt'/a(t')$.

For later application to CMB temperature fluctuation, we slightly generalized the line element to distingush the Gravitational potential $\Psi$ and potential for spatial curvature $\Phi$. The difference between two potentials appear when there are nondiagonal components in the energy momentum tensor. In this book, we need to distinguish these potential only when we apply the geodesic equation to CMB temperature fluctuation.

In general two metric tensors are said to be conformally related if they are related with an arbitrary function of spacetime $\Omega(x)$ as follows.

$$g_{\mu\nu} = \Omega^2(x)\tilde{g}_{\mu\nu} \tag{2.5.11}$$

In this case, Christoffel symbol calculated by these metric tensors are related as follows.

$$\tilde{\Gamma}^\mu_{\nu\rho}(\tilde{g}) = \Gamma^\mu_{\nu\rho}(g) + \delta^\mu_\nu\partial_\rho\ln\Omega + \delta^\mu_\rho\partial_\nu\ln\Omega - g^{\mu\lambda}g_{\nu\rho}\partial_\lambda\ln\Omega \tag{2.5.12}$$

Then for $k^\mu$ satisfying a null geodesic equation, we have

$$k^\nu \tilde{\nabla}_\nu k^\mu = k^\mu \partial_\nu k^\mu + k^\nu \tilde{\Gamma}^\mu_{\nu\rho} k^\rho = 2 \left( k^\nu \partial_\nu \ln \Omega \right) k^\mu \qquad (2.5.13)$$

Then one can show that the vector $\tilde{k}_\mu = \Omega^{-2} k_\mu$ satsifies the null geodesic equation in the metric $\tilde{g}_{\mu\nu}$.

$$\tilde{k}^\nu \tilde{\nabla}_\nu \tilde{k}_\mu = 0, \quad \tilde{g}_{\mu\nu} \tilde{k}^\mu \tilde{k}^\nu = 0 \qquad (2.5.14)$$

Let us write down the geodesic equation explicitly for the line element given above. Non-zero Christofel symbols are as follows for the line element (2.5.10) are as follows.

$$\Gamma^0_{\ 00} = \Psi', \quad \Gamma^0_{\ 0i} = \Psi_{,i}, \quad \Gamma^0_{\ ij} = -\Phi'\delta_{ij} \qquad (2.5.15)$$
$$\Gamma^i_{\ 00} = \Psi_i, \quad \Gamma^i_{\ 0j} = \Phi'\delta^i_j, \quad \Gamma^i_{\ jk} = -\Phi_{,j}\delta^i_k - \Phi_{,k}\delta^i_j + \Phi^{,i}\delta_{ij}$$
$$(2.5.16)$$

Here, the prime is the derivative with respect to the conformal time $\eta$.

## Time component of geodesic equation and CMB temperature flucutation

First, we apply the time component of the geodesic equation to CMB temperature fluctuation. From now on, we will no longer use tildes above the photon momentum for notational simplicity. The time component can be calculated to be

$$\frac{dk^0}{d\eta} + k^0 \left( \Psi' + 2\Psi_{,i} n^i - \Phi' \right) = 0 \qquad (2.5.17)$$

Here, $\Psi$ and $\Phi$ are considered to be infinitesimal quantities, so in an approximation that ignores the second-order infinitesimal quantities, the four-dimensional momentum of the photon multiplied to them can be taken as the background momentum $k^\mu = (k^0, k^0 n^i)$ where $n^i$ is a component of a 3-dimensional unit vector. Now, we write the above equation in the following form.

$$\frac{dk^0}{d\eta} = \left( \Psi' + \Phi' \right) \bar{k}^0 - 2 \left( \Psi' + n^i \partial_i \Psi \right) k^0 = 0 \qquad (2.5.18)$$

Then the second term in the right-hand side is the total derivative along the geodeisc and thus integrating from last scattering surface to the present gives

$$\frac{k^0(\eta_0) - k^0(\eta_{\rm LS})}{k^0(\eta_{\rm LS})} = \int_{\eta_{\rm LS}}^{\eta_0} d\eta' \left(\Psi' + \Phi'\right) - 2\left[\Psi(\eta_0) - \Psi(\eta_{\rm LS})\right]$$

(2.5.19)

where we have expanded the exponential function and ignored terms higher than the second orders. We further rewrite this equation as follows.

$$\frac{\omega(\vec{n}, \eta_0) - \omega(\vec{n}, \eta_{\rm LS})}{\omega(\vec{n}, \eta_{\rm LS})} \simeq \int_{\eta_{\rm LS}}^{\eta_0} d\eta' \left(\Psi' + \Phi'\right)$$
$$- \left[\Psi(\eta_0) - \Psi(\eta_{\rm LS}) + n^i \left(v^i(\eta_{\rm LS}) - v^i(\eta_0)\right)\right]$$

(2.5.20)

where we have used the following expression for the energy (frequency) of a photon $\vec{k}$ observed by the observer with 4 velocity $\vec{U}$

$$\omega(\eta) = -g_{\mu\nu} U^\mu k^\nu = \left[1 + \Psi(\eta) - n^i v^i(\eta)\right] k^0(\eta) \qquad (2.5.21)$$

Since the radiation energy density is proportional to the forth power of the temperature, we have

$$\delta_\gamma = \frac{\delta\rho_\gamma}{\rho_\gamma} = 4\frac{\delta T}{T} \qquad (2.5.22)$$

Finally, we have the following expression for the observed CMB temperature fluctuation.

$$\frac{\delta T}{T} = \frac{1}{4}\delta\rho_\gamma(\eta_{\rm LS}) + \Psi(\eta_{\rm LS}) + \int_{\eta_{\rm LS}}^{\eta_0} d\eta' \left(\Psi' + \Phi'\right) + n^i v^i(\eta_{\rm LS})$$

(2.5.23)

The first term on the right side is the temperature fluctuation due to the density fluctuation existing on the last scattering surface, and the higher the density, the higher the temperature will be than the average value. On the other hand, in regions with high density, the gravitational potential $\Psi < 0$ is also deep. The second term represents the decrease in temperature due to gravitational redshift.

The observed temperature fluctuation is the sum of the first and second terms, and is called Sacks–Wolfe effect.

There is a simple argument to estimate the Sack–Wolf effect according to W. Hu and M. White. If we consider the first term to be the time shift $\delta t/t = \Psi$ due to the gravitational potential $\Psi$ and use the expansion law $a \propto t^{2/3}$, then we get $\delta a/a = (2/3)\delta t/t$. Since temperature is inversely proportional to the scale factor, thus, $\delta T/T|_{LS} = -\delta a/a = -(2/3)\Psi$, In the end, the person who gave these two terms is

$$\left(\frac{\delta T}{T}\right)_{LS} = \frac{1}{3}\Psi \tag{2.5.24}$$

Therefore, we observe regions with higher density to have lower temperature.

The third integral term is called the integrated Sacks–Wolfe effect, and appears when there is a time change in the potential. Since the potential is constant when matter is dominant, the effect appears in the region near the last scattering surface where the influence of radiation remains and at the time when dark energy becomes the dominant energy of the universe. The last term is the Doppler effect due to the movement of electrons on the final scattering surface. Note that the equation (2.5.23) ignores the potential at the observation point and the Doppler effect since the effect of the potential disappears after averaging over all direction and the Doppler effect gives us the dipole anisotropy of the observed CMB temperature map which comes from the motion of the solar system with respect to the frame where CMB looks isotropic on average. and nothing to do with the anisotropy associated with the early universe.

## 2.6. Gravitational Lens Equation

Next, we derive gravitaional lens equation from geodesic equation. Since we are concerned with the period long after the last scattering and thus safely neglect the non-diagonal part of energy momentum tensor and set $\Psi = \Phi$. Then the time component of the geodesic equation reads

$$\frac{d^2\eta}{d\lambda^2} + 2\Psi_{,i}n^i\left(\frac{d\eta}{d\lambda}\right)^2 = 0 \tag{2.6.1}$$

where we have ignored terms higher than the second orders. The spatial component of the geodesic equation is

$$\frac{d^2x^i}{d\lambda^2} + \left(-2\Psi'n^i + 2(\delta^{ij} - n^in^j)\partial_j\Psi\right)\left(\frac{d\eta}{d\lambda}\right)^2 = 0 \tag{2.6.2}$$

These two equations give the following equation for the world line of a photon in an inhomogeneous universe.

$$\frac{d^2x^i}{d\eta^2} - 2\frac{d\Psi}{d\eta}n^i + 2\left(\delta^{ij} - n^in^j\right)\partial_j\Psi = 0 \tag{2.6.3}$$

For example, if a ray is initially traveling in the positive direction of the $z$ axis, $\vec{n} = (0,0,1)$, and the direction of travel is bent by the gravitational potential, but the change is very small, we will ignore the change in each component of the three-dimensional vector $\vec{n}$ (Born approximation). Therefore, in the approximation that ignores the second-order infinitesimal quantity,

$$\frac{d^2x^a}{d\eta^2} + 2\partial_a\Psi = 0, \quad a = 1,2 \tag{2.6.4}$$

$$\frac{d^2x^3}{d\eta^2} - 2\frac{d\Psi}{d\eta} = 0 \tag{2.6.5}$$

In the application to gravitational lenses described in the next section, it is more convenient to use the polar coordinates $(r, \theta, \phi)$ as the spatial coordinates and we are at the origin $r = 0$. Furthermore, the use of the polar coordinates allows us to handle the bending of light in general spaces with curvature, and thus we write the spatial line element as follows.

$$d\ell^2 = (1 - 2\Psi)\left(d\chi^2 + r^2(\chi)(d\theta^2 + \sin^2\theta d\phi^2)\right) \tag{2.6.6}$$

Then Christoffel symbols are calculated as

$$\Gamma^i{}_{jk} = -\Psi_{,j}\delta^i_k - \Psi_{,k}\delta^i_j + \Psi^{,i}g_{ij} + {}^{(3)}\Gamma^i{}_{jk} \qquad (2.6.7)$$

where ${}^{(3)}\Gamma^i{}_{jk}$ is the intrinsic Christoffel symbol for the averaged line element of (2.6.6) with $\Psi = 0$. (Note that Christoffell symbols does not vanish because we used curved coordinates even flat space). Thus, our equation becomes as follows

$$\frac{d^2x^i}{d\eta^2} - 2\frac{d\Psi}{d\eta}n^i + 2\left({}^{(3)}g^{ij} - n^in^j\right)\partial_j\Psi + {}^{(3)}\Gamma^i{}_{jk}n^jn^k = 0 \qquad (2.6.8)$$

where ${}^{(3)}g_{ij}$ is the metric tensor for the averaged spatial line element (2.6.6) without $\Psi$.

Now, if there is no inhomogeneities in the material distribution, the light propagates on a radial worldline where its spatial direction is given by the unit vector $\bar{n} = (1,0,0)$. In reality, due to inhomogeneous gravitational field, the path bends and $\theta$ and $\phi$ components appear, but these are of the order of the gravitational potential $\Phi$ and thus they are first order small quantities(that means the same order of the gravitational potential). Noting that ${}^{(3)}\Gamma^i{}_{jk}$ are also first order quantities, the following equations are obtained.

$$\frac{d^2\chi}{d\eta^2} - 2\frac{d\Psi}{d\eta} = 0 \qquad (2.6.9)$$

$$\frac{d^2\theta}{d\eta^2} + \frac{2}{r^2(\chi)}\frac{\partial\Psi}{\partial\theta} + 2\frac{d\ln r}{d\chi}\frac{d\theta}{d\eta} = 0 \qquad (2.6.10)$$

$$\frac{d^2\phi}{d\eta^2} + \frac{2}{r^2(\chi)\sin^2\theta}\frac{\partial\Psi}{\partial\phi} + 2\frac{d\ln r}{d\chi}\frac{d\phi}{d\eta} = 0 \qquad (2.6.11)$$

These equations are straightforwadly integrated. We show how to integrate the equation for $\theta$ as an example. The equation for $\theta$ can be written as

$$\frac{d}{d\eta}\left(r^2\frac{d\theta}{d\eta}\right) = -2\frac{\partial\Phi}{\partial\theta} \qquad (2.6.12)$$

Then for the case of $K = +1$, we have

$$\int_{\eta_0}^{\eta} d\eta' \frac{1}{r^2(\eta')} \int_{\eta_0}^{\eta'} d\eta'' \partial_\theta \Phi(\eta')$$

$$= \int_{\eta_0}^{\eta} \frac{1}{\sin^2(\eta_0 - \eta')} \int_{\eta_0}^{\eta'} d\eta' \partial_\theta \Phi(\eta')$$

$$= \frac{\cos(\eta_0 - \eta')}{\sin(\eta_0 - \eta')} \int_{\eta_0}^{\eta'} d\eta'' \partial_\theta \Phi(\eta'') \Big|_{\eta_0}^{\eta} - \int_{\eta_0}^{\eta} d\eta' \frac{\cos(\eta_0 - \eta')}{\sin(\eta_0 - \eta')} \partial_\theta \Phi(\eta')$$

$$= \int_{\eta_0}^{\eta} d\eta' \frac{\sin(\eta_0 - \eta') \cos(\eta_0 - \eta) - \cos(\eta_0 - \eta') \sin(\eta_0 - \eta)}{\sin(\eta_0 - \eta) \sin(\eta_0 - \eta')} \partial_\theta \Phi(\eta')$$

$$= \int_0^{\chi} d\chi' \frac{r(\chi - \chi')}{r(\chi) r(\chi')} \partial_\theta \Phi(\chi')$$

The cases $K = 0, -1$ are similarly calculated to have the following results.

$$\chi = \eta_0 - \eta - 2 \int_{\eta_0}^{\eta} \Psi d\eta' \tag{2.6.13}$$

$$\theta(\chi) = \theta_0 - 2 \int_0^{\chi} d\chi' \frac{r(\chi_s - \chi')}{r(\chi_s) r(\chi')} \frac{\partial}{\partial \theta} \Psi(\chi \hat{n}, \eta_0 - \chi') \tag{2.6.14}$$

$$\phi(\chi) = \theta_0 - 2 \int_0^{\chi} d\chi' \frac{r(\chi_s - \chi')}{r(\chi_s) r(\chi')} \frac{1}{\sin^2 \theta} \frac{\partial}{\partial \phi} \Psi(\chi \hat{n}, \eta_0 - \chi') \tag{2.6.15}$$

Here, we set $\chi = 0$ at $\eta = \eta_0$., and thus $\hat{n} = (\theta_0, \phi_0)$ is the direction of the observed source, and $(\theta(\chi), \phi(\chi))$ is the direction of light ray at the comoving distance $\chi$.

If we define the angle of bending of the ray as $\vec{\alpha} = (\theta_0 - \theta, \sin\theta_0(\phi_0 - \phi))$, then this angle is about 1 second for a galaxy, and about 10 seconds for a galaxy cluster. Therefore, it can be considered that the change in the direction of the light ray due to the gravitational lens occurs within a very small region on sky around the original direction of the light source (that is the direction to the source without lensing). If we introduce a 2-dimensional orthogonal coordinate system $(\theta_1, \theta_2) = (\theta \cos \phi, \theta \sin \phi)$ in a small region of the celestial sphere where we may regard the region is a flat 2-dimensional plane with the origin at the lensing object, the differentiation on this plane is

$$\nabla_\theta \equiv \left( \frac{\partial}{\partial \theta}, \frac{1}{\sin \theta} \frac{\partial}{\partial \phi} \right) = \left( \frac{\partial}{\partial \theta_1}, \frac{\partial}{\partial \theta_2} \right) \tag{2.6.16}$$

Then the bending angle can be expressed as follows.

$$\vec{\alpha} = \nabla_\theta \psi \tag{2.6.17}$$

Here, we have defined the lens potential as follows.

$$\psi(\vec{\theta}) = \int_0^{\chi_s} d\chi \frac{r(\chi_s - \chi)}{r(\chi) r(\chi_s)} \Psi(x^\mu(\chi)) \tag{2.6.18}$$

Here, the light source is assumed to be at $\chi = \chi_s$. $\vec{\theta}$ is the direction in which the ray comes. Strictly speaking, this integral should be performed along the world line $x^\mu(\chi)$ of the ray, but in reality, since the gravitational potential is very small, the integral is evaluated along the null rays connecting the observer and the light source in an homogeneous and isotropic background space-time.

In the end, we have the following gravitational lens equation

$$\vec{\beta} = \vec{\theta} - \vec{\alpha} \tag{2.6.19}$$

Here, all vectors are two-dimensional vectors on the region defined above, and $\vec{\beta}$ is a vector from the origin to the position of the unlensed source.

# Chapter 3

# Gravitational Lens

## 3.1. Introduction and History

Propagation of light rays from a distant source to the observer is governed by the gravitational field of the intervening mass distribution. This fact was realized even before Einstein formulated the general theory of relativity. It seems that Cavendish and Michell are the first who presented the idea that the bending of a light ray occurs near a massive body around 1780 (see Valls-Gabaud [1] for the detailed history). Then German astronomer, Soldner, calculated the bending angle of a light ray around a massive object in 1801 using Newtonian gravity and obtained one half of the correct value [2].

The correct value can be only calculated by knowing the fact that spacetime is curved by a massive object. General relativity provides a method to calculate the spatial curvature around a massive object. Einstein calculated the bending angle of a light ray grazing the surface of the sun correctly as 1.76 arcsec in 1915. In 1919, a team led by A. Eddington succeeded to measure the change in position of stars around the sun during an eclipse and obtained results consistent with the prediction by general relativity.

In 1924, the Russian scientist Chwolson pointed out the possibility of producing fake double stars by gravitational lensing and furthermore without detailed calculation he pointed out a possible existence of perfect ring image if the source is located exactly on the line of sight of the lens [4]. In 1936, a Czech amateur scientist, Mandl, visited Einstein at Princeton asking about the possible existence of multiple images of a distant star due to the gravitational force of a star. Stimulated by the discussion, Einstein wrote a paper on gravitational lensing by a star with detailed calculation and showed the existence of the so-called Einstein ring, predicted by Chwolson [5]. He concluded however, that gravitational lensing by a star is unlikely to occur because of the extremely small probability of lensing. It should be noted that Link studied similar lensing event before Einstein. He calculated not only the position of the image but also magnification [6, 7]. Zwicky estimated the probability of lensing of one galaxy by another galaxy as $\sim 10^{-3}$ (Ref. [8]) which is much larger than that of a star and tried to discover lensing phenomena in galaxy clusters. The attempt failed mainly because the lensed images are too faint to be detected by the apparatus available at that time.

After that there were no active researches on gravitational lensing except some important papers. In 1960, Russian astronomers Idlis and Gridneva wrote a paper which is regarded as the precursor of the weak lensing idea. In 1964, Refsdal [9] and Liebes [10] argued that the gravitational lensing effect may be used to measure the mass of a lensing object. Refsdal also pointed out that if a source's luminosity varies in time, there will be a time lag of variability between the lensed images which can be used to estimate the Hubble parameter.

The first lensing system QSO 0957+561A, B was discovered in early 1979 by Walsh, Carswell, and Weymann [11]. They made photometric observations of two images of a single quasar at $z = 1.41$. The lensing galaxy was also observed at redshift $z = 0.355$ in a cluster of galaxies. In 1997, the time lag of the variability was observed as $417 \pm 3$ days between images A and B [13]. Since then, other systems with multiple images have been observed, and the list has grown to several hundred by now. Time lags have been observed in mcre than 10 lensed systems as well. This list is expected to be on the order of 100 in the coming decade.

In 1980's the giant arcs in two clusters (Abell 370 and CL 2244-02) and pieces of elongated images in another cluster (Abell 2218) were discoved [14–16]. A multitude of arcs and distorted images in massive clusters have been described by Soucail *et al.* [16]. In the late 1980s, high resolution observations by the Hubble Space Telescope (HST) became available and showed that arcs are a universal feature of clusters. An Einstein ring-like image was also found in the radio source MG 1131+045 1988 using the Very Large Array [18]. Later HST discoverd many Einstein rings.

Gravitational lensing is classified into three regimes: strong, weak and micro lensing. Figure 3.1 shows a schematic representation

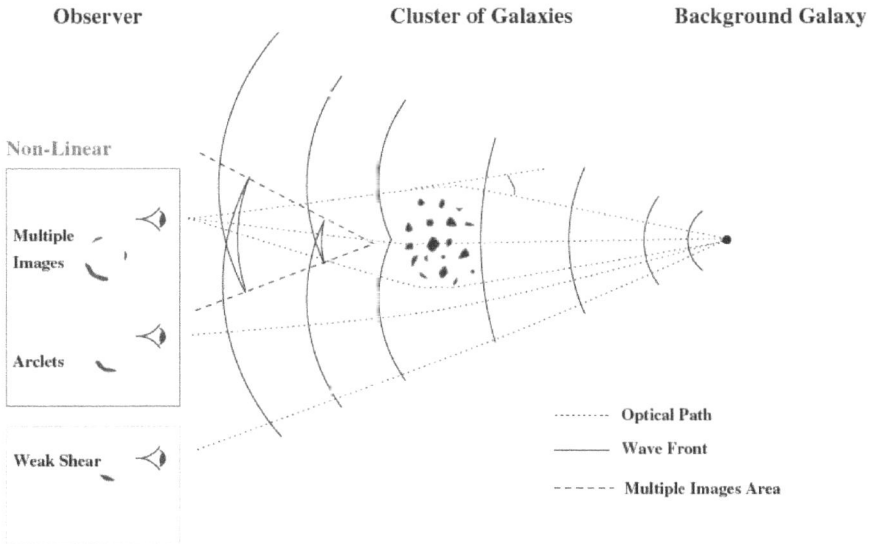

Fig. 3.1. The illustration of the gravitational lensing mechanism [?].

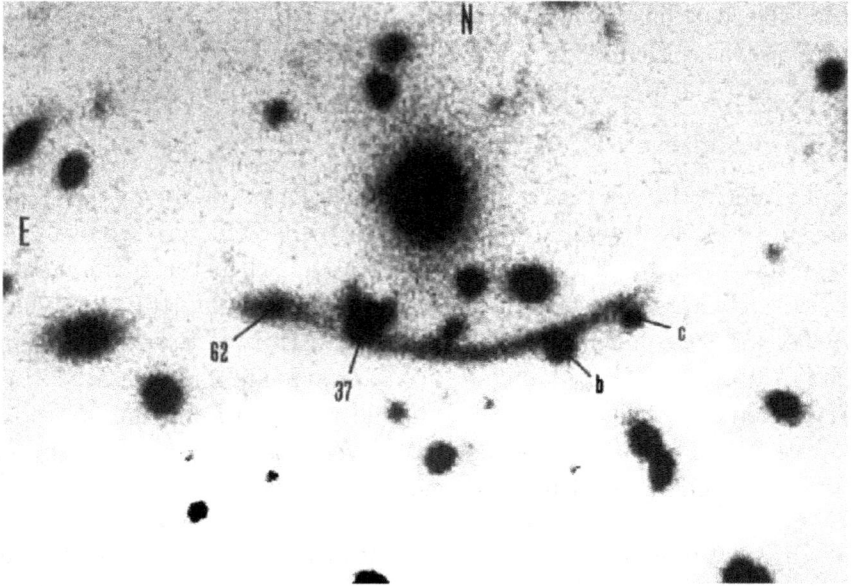

Fig. 3.2.  Giant arc in Abell 370 by Soucail *et al.* [16].

of strong and weak lensing phenomena. Strong lensing results in multiple images and/or highly deformed arclike images, called giant luminous arcs. It is used not only to measure the spatial dark matter distribution in the lensing object (e.g. Kneib *et al.* [19]), but also to constrain the cosmological parameters. It has been shown that a good reconstruction of a lensing system may constrain the density parameter and the cosmological constant (Futamase and Yoshida, 1999; Yamamoto *et al.*, 2001; Soucail *et al.*, 2004; Meneghetti *et al.*, 2005; Jullo *et al.* [24]).

An important application of strong lensing is to use it as a natural telescope. Strong lensing magnifies small and faint objects to make them observable, objects that are otherwise not observable. In particular, strong lensing by clusters has been used to study distant galaxies (Ellis *et al.* [25]; Kneib *et al.* [26]; Richard *et al.* [27]; Bouwens *et al.* [28]; Coe *et al.* [29]). For this study, a highly accurate determination of mass distribution is necessary, and this becomes possible using a large number of multiple-image and arcs discovered by the HST profitting from progress in lens-modeling techniques. It is claimed that the mass in cluster cores is measured to a few percent accuracy (e.g. Bradac *et al.* [30, 31]; Jullo *et al.* [32]; Jullo and Kneib [33]). In the case of clusters observed by the HST Frontier Fields, galaxies at $z > 10$ and as faint as 32 AB magnitude are expected to be detected (Richard *et al.* [34]). Future prospects for this study seem very promising.

Another application of strong lensing is lens statistics (Turner, Ostriker and Gott [35]). It has been shown that lens statistics are a good indicator of cosmological parameters, in particular, the cosmological constant (Turner [36]; Fukugita, Futamase and Kasai [37]).

Weak lensing results only in weakly deformed images and the lensing signal is obtained only by averaging over a certain number of background galaxies. While weak lensing is not spectacular compared with strong lensing, it is very important and useful. For example, strong lensing images are only observed in the central part of a cluster, which constitutes only a tiny part of the whole mass distribution of the cluster. Information from weak lensing is necessary to reveal the whole mass distribution of the cluster. The method of mass reconstruction by measuring the weak lensing effect was developed and applied to Abell 1689 and Cl 1409+53 by Tyson *et al.* [38]. More sophisticated methods of weak lensing mass reconstruction were developed by Kaiser and Squires [40] and Kaiser, Squires and Broadhurst (now known as the KSB method [41]). This method and its refinements have been applied to many clusters of galaxies. Weak lensing by large scale structure known as cosmic shear depends on the growth rate of the structure and thus on the cosmic expansion rate as well as the law of gravity. Thus, the accurate measurement of cosmic shear is expected to constrain the nature of dark energy and the deviation from general relativity. The first detection of cosmic shear was made in 2000 by several groups [43–48], and is one of the most active area of research in gravitational lensing.

Microlensing is a type of strong lensing by a stellar size object but whose deflection angle is on the order of micro-arcsec. Thus, no separate images are observed, but, only the magnification of the image is observed. It is used to discover MAssive Compact Halo Objects (MACHOs) [49] and planets outside of the solar system. We will not consider microlensing in this book.

For general review of gravitational lensing, we refer the reader to Schneider, Ehlers and Falco [50], Blandford and Narayan [51], Refsdal and Surdej [52] and Narayan and Bartelmann [53].

## 3.2. Basic Properties for Lens Equation

The global geometry and overall evolution of the universe is well described as homogeneous and isotropic by the Friedman–Robertson–Walker (FRW) metric. However, the matter distribution is highly inhomogeneous, up to 150 Mpc as we observe. A light ray propagates through these inhomogeneities feeling the local gravitational potential. How the averaged FRW universe appears from the local inhomogeneous universe within the framework of general relativity is a complicated and difficult problem, see for example Kasai and Futamase [39]. Fortunately, the gravitational lensing phenomena treated in this book are caused by weak gravitational fields and we may regard the field as locally stationary on the time scale of the deflection of a light ray. We have derived the lensing equation in such a situation. We start from the lensing equation derived in chapter 2.

$$\vec{\beta} = \vec{\theta} - \vec{\alpha} \qquad (2.6.19)$$

In the following we use the bold face letter to denote the 2-dimensional vector, where $\boldsymbol{\beta}$ is angular coordinate vector in the source plane and $\boldsymbol{\theta}$ is angular coordinate vector in the image plane, and the projected bending angle is expressed by the Newtonian potential $\Psi$ as follows:

$$\boldsymbol{\alpha} = 2 \int_0^{\chi_s} d\lambda \frac{r(\lambda_s - \lambda)}{r(\lambda_s)} \nabla^{(3)} \Psi(x(\lambda)) \qquad (3.2.1)$$

where $\nabla^{(3)}$ is the 3-dimensional gradient operator and $\partial_x = \partial_\theta/r(\lambda)$. This suggests that the bending angle may be expressed by the 2-dimensional gradient $\nabla_\theta$ of a 2-dimensional potential $\psi(\theta_1, \theta_2)$ called the lens potential

$$\alpha = \nabla_\theta \psi \qquad (3.2.2)$$

where $\nabla_\theta = (\partial_{\theta_1}, \partial_{\theta_2})$.

## Thin lens approximation

In particular, if the lensing object lies within a sufficiently small region between $\lambda_L \pm \Delta\lambda (\Delta\lambda \ll \lambda_L)$ such as clusters and galaxies compared with the relevant cosmological distances, one can assume that the light ray is deflected at $\lambda = \lambda_L$. The projected bending angle takes the form

$$\alpha = 2\frac{D_{LS}}{D_S} \int_{\lambda_L-\Delta\lambda}^{\lambda_L+\Delta\lambda} d\lambda \, \nabla^{(3)} \Psi(x(\lambda)) \qquad (3.2.3)$$

where $D_S = a(\lambda_s)r(\lambda_s)$ and $D_{LS} = a(\lambda_s)r(\lambda_s - \lambda_L)$ are the angular diameter distances from the observer to the source, and from the lensing object to the source, respectively. This is called the thin lens approximation. Figure 3 3 shows the configuration of the lensing equation in the thin lens approximation.

Using the above expression for $\alpha$ and Poisson equation $\Delta\Psi = 4\pi G\rho$, we have

$$2\kappa \equiv \nabla_\theta \cdot \alpha = \Delta_\theta \psi = \frac{D_L D_{LS}}{D_S} \frac{8\pi G\Sigma}{c^2} \qquad (3.2.4)$$

where $\Sigma = \int_0^{\chi_S} d\chi \rho(\chi)$ is the surface mass density along the line of sight.

We also defined the convergence $\kappa$, which is the normalized surface mass density

$$\kappa(\theta) \equiv \frac{\Sigma(\theta)}{\Sigma_{\mathrm{cr}}} \qquad (3.2.5)$$

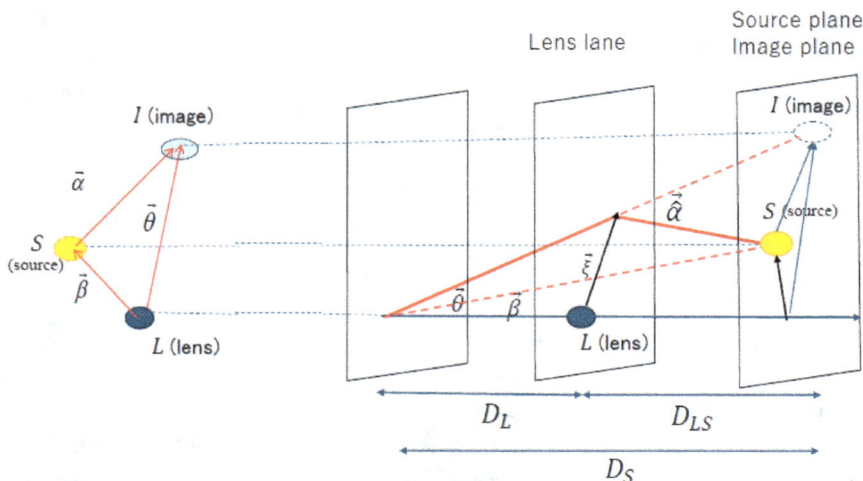

Fig. 3.3.  Geometry of the gravitaional lensing in thin lens approximation.

where the critical surface mass density is given by

$$\Sigma_{\rm cr} \equiv \frac{c^2 D_S}{4\pi G D_{LS} D_L} \approx 2.45 \times 10^{-1} h \left(\frac{d_{LS} d_L / d_S}{1\,{\rm Gpc}}\right)^{-1} {\rm g/cm}^{-2}$$

$$(3.2.6)$$

where $d_X$ denotes the angular diameter distance $D_X$ in units of $c/H_0$.

A region of a mass distribution with $\Sigma > \Sigma_{\rm cr}$ can produce multiple images for some source positions and is referred to as being super-critical. One can write the bending angle in the following more tractable form using the surface mass density $\Sigma$, or the convergence $\kappa$:

$$\alpha = \frac{4G}{c^2} \frac{D_L D_{LS}}{D_S} \int d^2\theta' \Sigma(\boldsymbol{\theta}') \frac{\boldsymbol{\theta} - \boldsymbol{\theta}'}{|\boldsymbol{\theta} - \boldsymbol{\theta}'|^2} = \frac{1}{\pi} \int d^2\theta' \kappa(\boldsymbol{\theta}') \frac{\boldsymbol{\theta} - \boldsymbol{\theta}'}{|\boldsymbol{\theta} - \boldsymbol{\theta}'|^2}$$

$$(3.2.7)$$

### 3.2.1.  *Properties of lens mapping*

One may regard the lens equation as mapping from the source plane to the image plane. It is not a one-to-one mapping, in general. The

Jacobian of the mapping characterizes the local distortion induced by the mapping:

$$A_{ij} = \frac{\partial \beta_i}{\partial \theta_j} = \delta_{ij} - \alpha_{ij} = \delta_{ij} - \psi_{,ij} \tag{3.2.8}$$

This 2 by 2 symmetric matrix A may be decomposed into a trace and traceless part as follows:

$$A(\boldsymbol{\theta}) = \begin{pmatrix} 1 - \kappa - \gamma_1 & -\gamma_2 \\ -\gamma_2 & 1 - \kappa + \gamma_1 \end{pmatrix} = (1 - \kappa)\begin{pmatrix} 1 & 0 \\ 0 & 1 \end{pmatrix} - \begin{pmatrix} \gamma_1 & \gamma_2 \\ \gamma_2 & -\gamma_1 \end{pmatrix} \tag{3.2.9}$$

where $\gamma \equiv \gamma_1 + i\gamma_2$ is the gravitational complex shear defined by

$$\gamma_1 = |\gamma|\cos(2\phi) = \frac{1}{2}(\psi_{,11} - \psi_{,22}) \tag{3.2.10}$$

$$\gamma_2 = |\gamma|\sin(2\phi) = \psi_{,ij} \tag{3.2.11}$$

In general the quantity $a$ is called a spin-s quantity when $a$ changes as

$$a' = e^{is\phi}a \tag{3.2.12}$$

under the rotation of coordinates by an angle $\phi$. The convergence and gravitational shear are then the spin-0 and spin-2 quantities, respectively. Using this property, it is convenient to define the tangential and radial components of the shear relative to some reference point $\boldsymbol{\theta}_0$. Noticing that the tangential component of the shear is obtained from $\gamma$ by rotating the angle $\pi/2 - \phi$ (where $\phi$ is the polar angle of $\boldsymbol{\theta} - \boldsymbol{\theta}_c$)(see Fig. 3.4), we have the following expressions for the real part and imaginary part of the transformed shear.

$$\gamma_t(\boldsymbol{\theta}; \boldsymbol{\theta}_0) = -\text{Re}(\gamma e^{-2i\phi}) = -\gamma_1 \cos 2\phi + \gamma_2 \sin 2\phi \tag{3.2.13}$$

$$\gamma_\times(\boldsymbol{\theta}; \boldsymbol{\theta}_0) = -\text{Im}(\gamma e^{-2i\phi}) = \gamma_1 \sin 2\phi - \gamma_2 \cos 2\phi \tag{3.2.14}$$

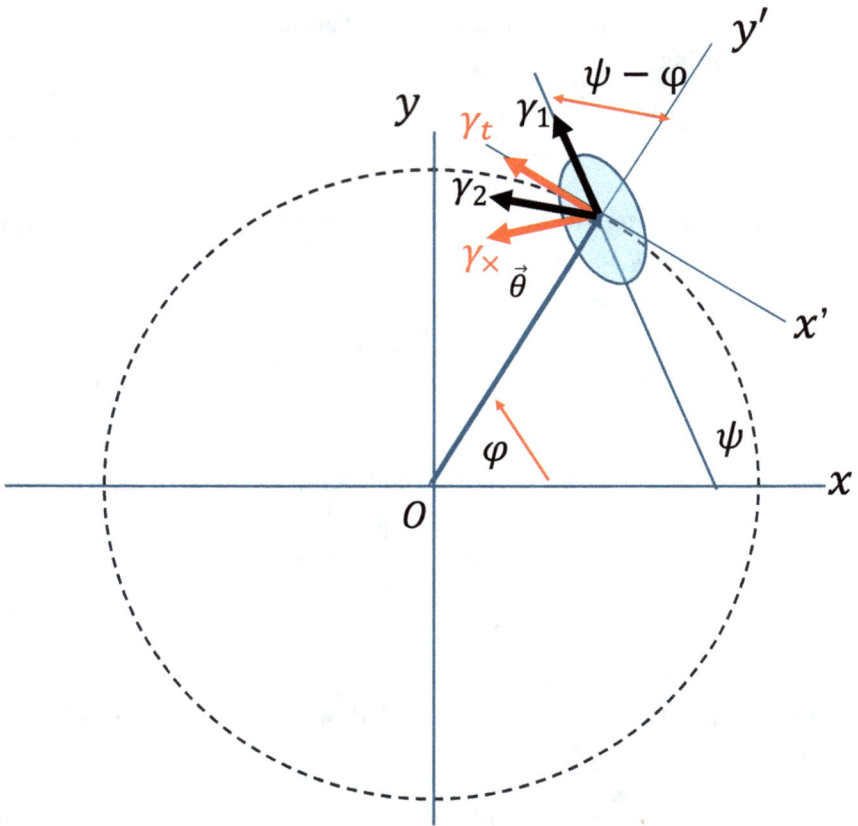

Fig. 3.4.  The configuration of the tangential shear.

$\gamma_t$ and $\gamma_\times$ are referred to as tangential and cross shears, respectively. Under the parity transformation, these change as $\gamma_t \to \gamma_t$, $\gamma_\times \to -\gamma_\times$. Tangential shear plays an important role in cluster mass reconstruction and cosmic shear, as shown later.

The convergence and the shear are related in the following way:

$$\gamma(\boldsymbol{\theta}) = \frac{1}{\pi} \int d^2\theta' D(\boldsymbol{\theta} - \boldsymbol{\theta}')\kappa(\boldsymbol{\theta}') \qquad (3.2.15)$$

with

$$D(\boldsymbol{\theta}) = \left[ \frac{1}{2} \left( \frac{\partial^2}{\partial \theta_1^2} - \frac{\partial^2}{\partial \theta_2^2} \right) + i \frac{\partial^2}{\partial \theta_1 \partial \theta_2} \right] \ln |\boldsymbol{\theta}| = \frac{\theta_2^2 - \theta_1^2 - 2i\theta_1\theta_2}{|\boldsymbol{\theta}|^4}$$

$$= -\frac{e^{2i\phi}}{\theta^2} \tag{3.2.16}$$

where we have introduced polar coordinates $(\theta, \phi)$ such that $\boldsymbol{\theta} = (\theta \cos \phi, \theta \sin \phi)$. Note that there exists a global transformation that leaves the shear $\gamma(\boldsymbol{\theta})$ invariant such that

$$\kappa(\boldsymbol{\theta}) \to \kappa(\boldsymbol{\theta}) + \kappa_0 \tag{3.2.17}$$

where $\kappa_0$ is an arbitrary constant. This degeneracy leads to an ambiguity in the mass reconstruction based only on the image distortion in weak lensing analysis.

In practical calculations, it is more convenient to work in Fourier space because one can make use of the Fast Fourier transform. The kernel is expressed as follows in Fourier space:

$$\hat{D}(k) = \pi \frac{k_1^2 - k_2^2 + 2ik_1k_2}{k^2} = \pi e^{2i\phi_k} \tag{3.2.18}$$

where $\boldsymbol{k} = (k, \phi_k)$ is the vector of the polar coordinates in Fourier space. Since $D(k)D^*(k) = \pi^2$, the inverse of (3.2.15) is straightforward in Fourier space:

$$\hat{\kappa}(k) = \frac{1}{\pi} D^*(k) \hat{\gamma}(k) \tag{3.2.19}$$

where $\hat{\kappa}$ and $\hat{\gamma}$ are the Fourier transforms of the corresponding quantities.

The geometrical meaning of convergence and shear is easily seen by considering the deformation of a circular source. Let us consider a circle with radius $d\beta$. Consider a small circle defined by the equation

$$\mathbf{d}\beta^T \mathbf{d}\beta = 1 \tag{3.2.20}$$

where $\mathbf{d}\beta$ is a position vector from the origin and $\mathbf{d}\beta^T$ is its transverse. Then the image mapped by the lensing satisfies the following equation:

$$1 = \mathbf{d}\theta^T A^T A \mathbf{d}\theta \tag{3.2.21}$$

The matrix $A$ has the following eigenvalues:

$$\lambda_\pm = 1 - \kappa \pm |\gamma| = (1 - \kappa)(1 \pm |g|) \qquad (3.2.22)$$

where $g = \gamma/(1 - \kappa)$ is called the reduced shear. The associated eigenvectors are

$$\mathbf{V}_+ = \begin{pmatrix} \cos\phi \\ -\sin\phi \end{pmatrix} = \begin{pmatrix} \cos\left(\phi - \frac{\pi}{2}\right) \\ \sin\left(\phi - \frac{\pi}{2}\right) \end{pmatrix}, \quad \mathbf{V}_- = \begin{pmatrix} \cos\phi \\ \sin\phi \end{pmatrix}$$

$$(3.2.23)$$

where $\phi = \frac{1}{2}\tan^{-1}(g_2/g_1)$ is the position angle of the eigenvector $\mathbf{V}_-$. Using the orthogonal matrix $O$ which diagonalizes the matrix $A$, one finds that the unit circle is mapped to the following ellipse (see Fig. 3.5)

$$\left(\frac{d\Theta_1}{\frac{1}{\lambda_-}}\right)^2 + \left(\frac{d\Theta_2}{\frac{1}{\lambda_+}}\right)^2 = 1 \qquad (3.2.24)$$

where $d\Theta = Od\theta$. The $d\Theta_1$ is the coordinate along the eigenvector $\mathbf{V}_-$.

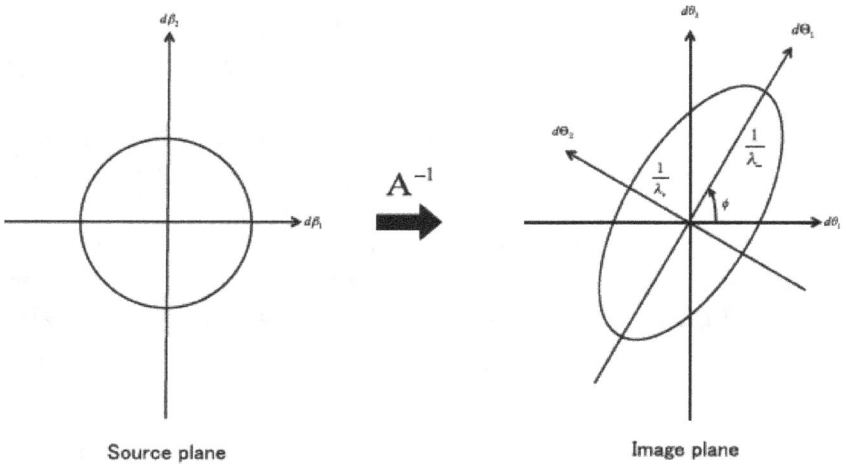

Source plane          Image plane

Fig. 3.5. The unit circle is mapped to an ellipse with the length of major axis $1/|\lambda_-|$ and minor axis $1/|\lambda_+|$ for $\kappa < 1$ and its inclination from $d\theta$ axis is $\phi$.

Since $1/|\lambda_-| > 1/|\lambda_+|$ for $\kappa < 1$, this is the equation of an ellipse elongated in the direction of $\phi$ (tangential direction) with the length of the major axis $1/|\lambda_-|$ and minor axis $|1/\lambda_+|$. For $\kappa > 1$, the ellipse is elongated in the direction $\phi - \pi/2$ (radial direction) with the length of the major axis $1/|\lambda_+|$ and the minor axis $1/|\lambda_-|$. It is interesting to note that the distortion disappears along the curve defined by $\kappa(\boldsymbol{\theta}) = 1$, which always lies in the odd-parity region. Moreover, in the region defined by $\kappa \simeq 1$ and $\gamma \simeq 0$ the mapped image is highly magnified without deformation. These are referred to as GRAavitationally lensed yet Morphologically Regular images, or GRAMORs. Such an image was predicted in 1998 by Williams and Lewis [55] and Futamese, Hattori and Hamana [56] and was discovered in the galaxy cluster MACS J1149.5+2223 (Broadhurst *et al.* [57]).

Thus, the convergence expresses an isotropic magnification ($\kappa > 0$) or demagnification ($\kappa < 0$). The (reduced) shear $g$ expresses a deformation without changing the area. The component $g_1$ expresses the deformation along the $\theta_1(\theta_2)$ axis for $g_1 > 0$ ($g_1 < 0$). The component $g_2$ expresses the deformation along the axis rotated 45° from the $\theta_1$ axis.

### 3.2.2. *Caustic and critical curves*

According to Liouville's theorem, surface brightness is conserved by gravitational lensing. This may be shown as follows. The energy in the interval $d\nu$ radiated in the solid angle $d\Omega$ through the area $ds$ in the time $dt$ may be written as $Icdtdsd\nu d\Omega$ ($I$ is the surface brightness). This may also be written using the photon distribution function $f(x,p)$ as

$$Icdtdsd\nu d\Omega d\nu = h\nu f d^3 x d^3 p = 4\pi h^3 \nu^3 f cdtdsd\nu d\Omega \quad (3.2.25)$$

Thus, $I \propto \nu^3 f$. Liouville's theorem tells us that $f$ = constant and $\nu$ = constant. Because there is no energy exchange in gravitational lensing, the surface brightness is conserved. Thus, the magnification of the flux due to lensing is given simply by ratio of the solid angle between image and source.

This is given by the inverse of the determinant of the Jacobian matrix $A$:

$$\mu = \frac{1}{\det A} = \frac{1}{(1-\kappa)^2(1-|g|^2)} = \frac{1}{(1-\kappa)^2 - |\gamma|^2} \quad (3.2.26)$$

Images with $\det A > 0$ ($\lambda_- > 0$ or $\lambda_+ < 0$) have the same parity as the source, while images with $\det A < 0$ ($\lambda_+ > 0$ and $\lambda_- < 0$) have parity opposite to the source.

The closed curves in the image plane on which the point image is magnified indefinitely are called "critical curves" which is defined by the condition $\lambda_\pm = 0$. The corresponding curves in the source plane are called "caustics". The critical curves separate the image plane into even- and odd-parity regions. In practice, realistic objects like galaxies have a finite size and $\det A = 0$ applies to every part of the object to be magnified indefinitely. Thus, the extended source is not magnified indefinitely on the critical curve, but is highly magnified. Such images are called luminous giant arcs. If the lensing object is sufficiently massive and satisfies $\kappa > 1$ in some region, then $\det A(\boldsymbol{\theta}) = 0$ is satisfied somewhere in the image plane and a critical curve exists. When convergence is a decreasing function of the distance from the center of the lensing object, the curve defined by $\lambda_+ = 0$ appears near the center and is called the inner critical curve and then the curve defined by $\lambda_- = 0$ appears outside and is called the outer critical curve. Since $\kappa > 1$ on the inner critical curve, the image is elongated in the radial direction (radial arc), and thus the inner critical curve is sometimes called the radial critical curve. On the other hand, $\kappa < 1$ on the outer critical curve, so the image on the outer critical curve is elongated in the tangential direction (tangential arc) and the outer critical curve is called the tangential critical curve.

The configuration of caustics determines the number of images. This is explained by the lens model examples below. The point determined by the condition $\kappa(\boldsymbol{\theta}) = 1$, namely $\Sigma(\boldsymbol{\theta}) = \Sigma_{cr}$ is the critical point in the absence of shear. Although the shear is non-zero in general, the region where $\kappa(\boldsymbol{\theta}) \geq 1$ is referred to as the strong lensing region associated with multiple images, while the region where $\kappa(\boldsymbol{\theta}) \ll 1$ is referred to as the weak lensing region.

## Circular lenses

Let us now consider simple, but very useful examples of a lens model. First, we consider a lens with a spherically symmetric surface mass distribution. In this case, the bending angle at a point $\boldsymbol{\theta}$ is determined by the projected mass $M(\theta)$ enclosed by a circle of radius $\theta = |\boldsymbol{\theta}|$ around the center of mass, and is given by

$$\boldsymbol{\alpha}(\boldsymbol{\theta}) = \frac{D_{LS}}{D_L D_S}\frac{4GM(\theta)}{c^2\theta^2}\boldsymbol{\theta} = \bar{\kappa}(\theta)\boldsymbol{\theta} \tag{3.2.27}$$

where $\bar{\kappa}$ is the average of $\kappa$ inside the radius $\theta$:

$$\bar{\kappa}(\theta) = \frac{1}{\Sigma_{cr}}\frac{M(\theta)}{\pi(D_L\theta)^2} = \frac{D_{LS}}{D_L D_S}\frac{4GM(\theta)}{c^2\theta^2} \tag{3.2.28}$$

The lens equation becomes the scalar equation:

$$\beta(\theta) = \theta - \bar{\kappa}(\theta)\theta \tag{3.2.29}$$

This shows immediately that the image with $\theta \geq 0$ satisfies $\theta \geq \beta$ and an image with $\theta < 0$ (opposite side of the source projected on the image plane) lies in the region with $\bar{\kappa}(\theta) > 1$. The Jacobian matrix is

$$A(\boldsymbol{\theta}) = \left(1 - \bar{\kappa} - \frac{\theta}{2}\frac{d\bar{\kappa}}{d\theta}\right)\begin{pmatrix} 1 & 0 \\ 0 & 1 \end{pmatrix}$$

$$-\frac{1}{2\theta}\frac{d\bar{\kappa}}{d\theta}\begin{pmatrix} (\theta_1)^2 - (\theta_2)^2 & 2\theta_1\theta_2 \\ 2\theta_1\theta_2 & -(\theta_1)^2 + (\theta_2)^2 \end{pmatrix} \tag{3.2.30}$$

Remembering that the mean convergence is a decreasing function of the distance from the center in general and $\gamma = -\theta/2 d\bar{\kappa}/d\theta$, the eigenvalues are then calculated to be

$$\lambda_+ = 1 - \kappa + \gamma = 1 - \frac{d}{d\theta}\left[\theta\bar{\kappa}(\theta)\right] = \frac{d\beta}{d\theta} \tag{3.2.31}$$

$$\lambda_- = 1 - \kappa - \gamma = 1 - \bar{\kappa}(\theta) = \frac{\beta}{\theta} \tag{3.2.32}$$

As explained above, the eigenvalue $\lambda_-$ describes the tangential deformation of the image due to a geometrical effect, and $\lambda_+$ describes the radial deformation of the image due to the gravitational tidal effect which depends on the radial density profile of the lensing object.

In the spherical lens case, the lens mapping may be understood by specifying a curve in the $(\beta, \theta)$ plane. The curve is characterized by a mass distribution. Figure 3.6 shows the mapping given by a

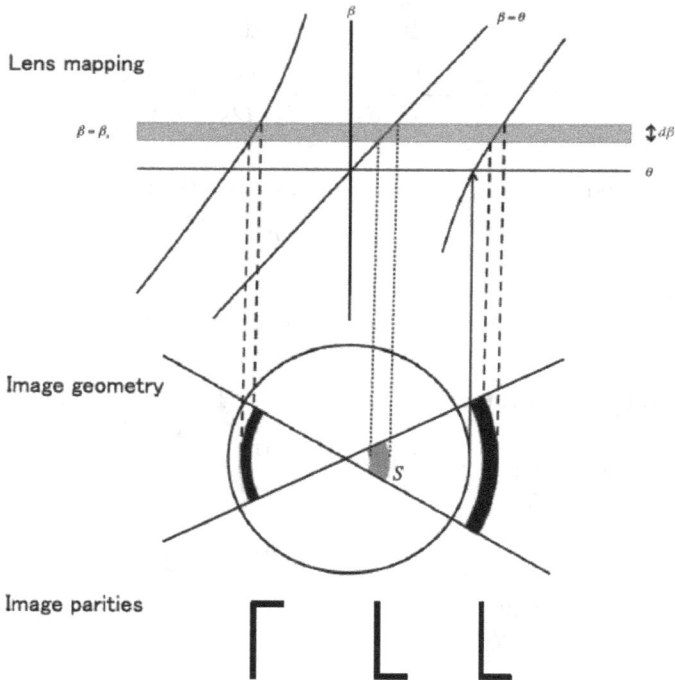

Fig. 3.6. Lens mapping given by a point mass.

**Lens mapping**

**Image geometry**

**Image parities**

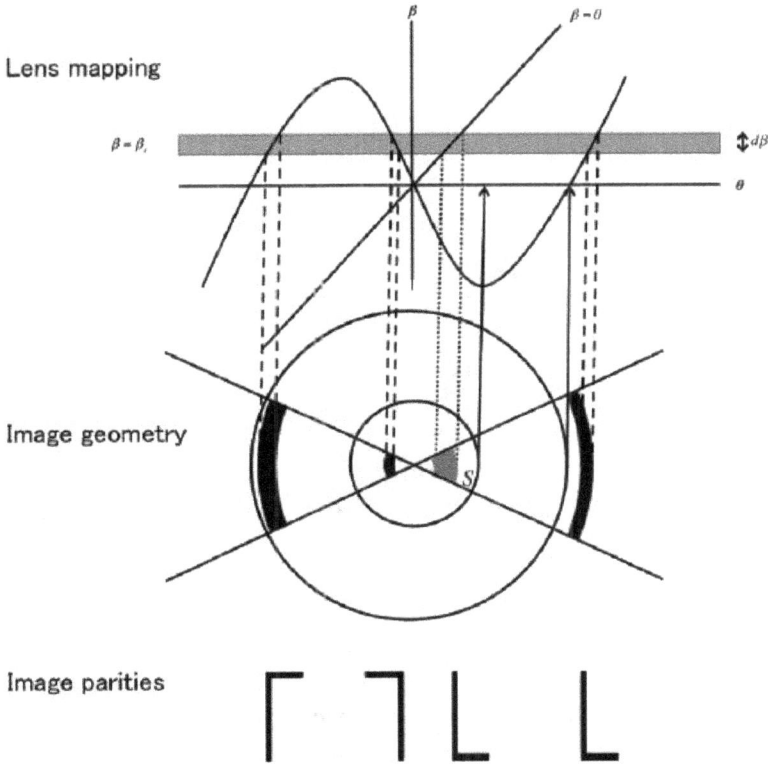

Fig. 3.7. Lens mapping given by a non-singular lens model.

point mass where there are only two images. If the lensing mass distribution is non-singular at the origin, the mapping is shown in Fig. 3.7. As seen in this figure, the caustic corresponding to the outer critical curve is the origin in the source plane (actually the line in this case).

When a source of a finite size approaches the caustic from the inside, two images with opposite parity approach perpendicularly to the inner critical line and these two images disappear when the source passes across the caustics. The opposite phenomenon occurs when the source approaches the caustic from the outside. Thus, the source must be inside the caustics to make multiple images. This may

**Source Plane**         **Image Plane**

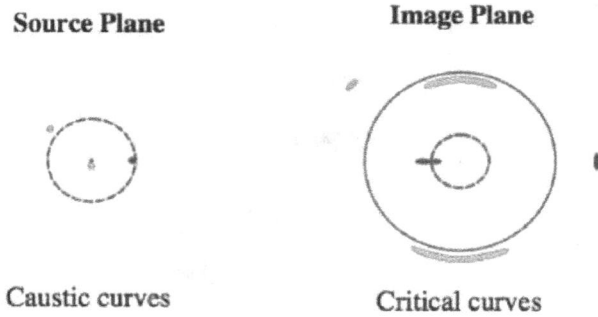

Caustic curves        Critical curves

Fig. 3.8. Causrics and critical curves for a spherical lens.

be more easily seen in Fig. 3.8, where the critical lines and caustics
are shown in the image plane and source plane, respectively. This is a
general property for caustics which are locally regarded as a straight
line. Such caustics are called "fold caustics".

**The Einstein radius and radial arcs**

From symmetry, the tangential caustics are given by $\beta = 0$. The
image is a ring on the tangential critical curve, provided that the lens
configuration is super-critical. Such a ring is known as an Einstein
ring and the angular radius is called the Einstein radius. The Einstein
radius is obtained by solving the lensing equation by taking $\beta = 0$:

$$\theta_E = \left[\frac{4GM(\theta_E)}{c^2}\frac{D_{LS}}{D_L D_S}\right]^{1/2}$$

$$= 29\,\text{arcsec}\left[\frac{M(\theta_E)}{10^{14}h^{-1}M_\odot}\right]^{1/2}\left(\frac{D_L D_S/D_{LS}}{1\,\text{Gpc}}\right)^{-1/2} \tag{3.2.33}$$

In actual lensing events, a source such as a galaxy is not perfectly
located at $\beta = 0$. In that case, we observe a pair of highly magnified,
stretched images (called arcs) along the tangential critical curve, with
a shorter arc just inside and a longer arc just outside of the critical
curve. Thus, the position $\theta_{\text{arc}}$ of the tangential arcs is used to roughly

estimate the mass of a lensing object inside the circle traced by the arc as follows:

$$M(\theta_{\text{arc}}) \simeq M(\theta_E) = 4.91 \times 10^{13} \left( \frac{D_L D_S / D_{LS}}{1 \, \text{Gpc}} \right) \left( \frac{\theta_E}{20''} \right)^2 M_\odot$$

(3.2.34)

Even for general non-circularly symmetric lenses, we define the effective Einstein radius as follows.

$$\theta_{,\text{eff}} = \sqrt{\frac{A_c}{\pi}}$$

(3.2.35)

where $A_c$ is the area inside the outer critical line. This definition reduces to the Einstein radius in circular symmetric case. The mass within the radius $\theta \sim \theta_{\text{eff}}$ around the effective Einstein angle is not influenced by assumption on the lens modeling.

$$M(\theta) = \Sigma_{cr} D_L^2 \int_{\theta < \theta_{\text{eff}}} d^2 \theta' \, \kappa(\theta')$$

(3.2.36)

Thus, the mass within the effective Einstein angle estimated in this way is a fundermental observable in the strong lensing region.

We now consider radial arcs which are sometimes observed in galaxy clusters. As shown above, the condition needed to have a radial arc is $\lambda_+ = 0$. Radial arcs are very useful to study the radial density profile. For example, let suppose that the convergence has a simple power law form as

$$\kappa(\theta) = \kappa_0 \left( \frac{\theta_0}{\theta} \right)^n$$

(3.2.37)

with $0 < n < 2$. The mean convergence is then given by

$$\bar{\kappa}(\theta) = \frac{\kappa(\theta)}{1 - \frac{n}{2}}$$

(3.2.38)

The condition $\lambda_+ = 0$, i.e. $d[\theta\bar{\kappa}]/d\theta = 0$ becomes

$$\theta_r = \left(\frac{1}{\kappa_0(1-n)}\right)^{1/n}\theta_0 \qquad (3.2.39)$$

Thus, the condition $n < 1$ is required for a lens to have a radial critical curve.

In circular lens, the lens mapping becomes one-dimensonal problem, and is explicitly obtained by specifying a curve in $(\beta, \theta)$ plane. The curve represents mass distribution in radial direction. We give two examples below.

**Point mass lens**

A point-mass model, in which mass is concentrated at one point, is used to describe gravitational lensing caused by a star, or by a supermassive black hole at the center of a galaxy.

When the lens with mass $M_P$ is located at the origin in the lens plane, the lens equation becomes

$$\beta = \theta - \frac{D_{LS}}{D_S}\frac{4GM_P}{c^2 D_L \theta} \qquad (3.2.40)$$

The Einstein radius is obtained by putting $\beta = 0$ as

$$\theta_E = \sqrt{\frac{4GM_P}{c^2}\frac{D_{LS}}{D_L D_S}} \qquad (3.2.41)$$

$$\simeq 0.9\,\mathrm{mas}\left(\frac{M_P}{M_\odot}\right)^{1/2}\left(\frac{D_L D_S/D_{LS}}{10\,\mathrm{kpc}}\right)^{-1/2} \qquad (3.2.42)$$

$$\simeq 0.9\,\mathrm{arcsec}\left(\frac{M_P}{10^{11}M_\odot}\right)^{1/2}\left(\frac{D_L D_S/D_{LS}}{10\,\mathrm{Gpc}}\right)^{-1/2} \qquad (3.2.43)$$

where mas means milli arcsecond. The lens equation has two solutions.

$$\theta_\pm = \frac{1}{2}\left(\beta \pm \sqrt{\beta^2 + \theta_E^2}\right) \tag{3.2.44}$$

Eigen values of Jacobi matrix is

$$\lambda_\pm = 1 \pm \left(\frac{\theta}{\theta_E}\right)^2 \tag{3.2.45}$$

Distortion in the tradial direction and tangential direction are given by

$$W = \frac{d\theta}{d\beta} = \frac{1}{1 + (\theta_E/\theta)^2}, \quad L = \frac{\theta}{\beta} = \frac{1}{1 - (\theta_E/\theta)^2} \tag{3.2.46}$$

The magnifications of each images are given by

$$\mu_\pm = \frac{1}{1 - (\theta_E/\theta)^4} - \frac{u^2 + 2}{2u\sqrt{u^2 + 4}} \pm \frac{1}{2} \tag{3.2.47}$$

where $u = \beta/\theta$. With a stellar-mass lens like MACHO, the separation angle between images is on the order of microseconds, and since they cannot be separated by optical observation, the two images are observed overlapping, and the overall magnification becomes as follows.

$$\mu = \mu_+ - \mu_- = \frac{u^2 + 2}{u\sqrt{u^2 + 4}} \tag{3.2.48}$$

When $u = 1$, $\mu = 1.34$, which means that it becomes 0.32 magnitude brighter.

From the bending angle $\alpha = \frac{\theta_E^2}{\theta}$, one can easily calculate the lens potential as $\psi(\theta) = \theta_E^2 \ln |\theta|$, The convergence and shear can be caculate from the lens potential

$$\kappa(\boldsymbol{\theta}) = \pi \theta_E^2 \delta_D^{(2)}(\boldsymbol{\theta}) \tag{3.2.49}$$

$$\gamma_1 = -\frac{\theta_E^2}{\theta^4}\left(\theta_1^2 - \theta_2^2\right) \tag{3.2.50}$$

$$\gamma_1 = -2\frac{\theta_E^2}{\theta^4}\theta_1\theta_2 \tag{3.2.51}$$

where $\delta_D^{(2)}$ is 2-dimentional Dirac delta function. The convergence vanishes everywhere except at the origin and can be expressed by 2-dimensional Dirac delta function. The averaged convergence and the tangential shear coincide

$$\bar{\kappa}(\theta) = \gamma_t(\theta) = \left(\frac{\theta_E}{\theta}\right)^2 \tag{3.2.52}$$

**Singular Isothermal Sphere**

A simple and useful model of a galaxy is a singular isotermal sphere which gives a flat rotation curve. The density is given by

$$\rho(r) = \frac{\sigma^2}{2\pi G r^2} \tag{3.2.53}$$

where $\sigma$ is the 1-dimensional velocity dispersion. The projected mass within an angular radius $\theta$ is thus given by

$$M(<\theta) = \frac{\pi\sigma^2}{G}D_L\theta \tag{3.2.54}$$

The bending angle is a constant given by the velocity dispersion.

$$\hat{\alpha} = 4\pi \left(\frac{\sigma}{c}\right)^2 \sim 1.4\,\text{arcsec}\left(\frac{\sigma}{220\,\text{km/s}}\right)^2 \tag{3.2.55}$$

The lens potential is

$$\psi(\theta) = \theta_E\theta \tag{3.2.56}$$

where $\theta_E = 4\pi \left(\frac{\sigma}{c}\right)^2 \frac{D_{LS}}{D_S}$ is the Einstein angle. Convergence and shear is given by

$$\kappa(\theta) = |\gamma(\theta)| = \frac{\theta_E}{2\theta} \tag{3.2.57}$$

The lens equation in this case becomes simply $\beta = \theta - \theta_E$, and there are two or one image depending on the position at the sorce. There are two images at $\theta_\pm = \theta_E \pm \beta$ for $|\beta| \le \theta_E$, otherwise only one image at $\theta = \theta_E + \beta$.

## NFW model

The next example is the so-called NFW profile, which is a numerically predicted universal profile for CDM halos given by Navarro, Frenk and White (NFW) [205]. Although it is known that this model does not correctly reproduce the dark matter halo density distribution beond the virial radius, the NFW lens model shown below has been found to reproduce well the observed results of gravitational lensing within and around haloes at the galaxy cluster scale.

The density of NFW model is given by

$$\rho(r) = \frac{\rho_s}{\left(\frac{r}{r_s}\right)\left(1 + \frac{r}{r_s}\right)^2} \tag{3.2.58}$$

where $\rho_c$ is the parameter corresponding to the central density. Sometimes it is expreesed by a dimensionless parameter $\delta_c$ defined by

$$\rho_c = \rho_{\rm crit}(z_L)\delta_c \tag{3.2.59}$$

where $\rho_{\rm crit}(z) = 3H^2(z)/(8\pi G)$ is the critical density at redshift $z$. Then $\delta_c$ is interpreted to represent the characteristic density fluctuation of the dark matter halo. $r_s$ is the parameter that expresses the characteristic scale where the density gradient satisfies $d\ln\rho(r)/d\ln r = -[1 + 3(r/r_s)]/[1 + (r/r_s)] = -2$.

Suppose that $r_\Delta$ represents the halo radius where the averaged mass density inside it is $\Delta$ times the critical density of the universe

$\rho_{\rm crit}$. In the spherically symmetric model, the density fluctuation that gravitationalyl collapses and reaches virial equilibrium in the Einstein–Dositter universe has an excess density $\Delta \simeq 18\pi^2 \sim 178$ at the time of formation, so $\Delta = 200$ is used in the definition of halo mass. The mass of NFW halo inside the radius $r_\Delta$ is given by

$$M_\Delta = \frac{4\pi\rho_s r_\Delta^3}{c_\Delta^3} \left[ \ln(1+c_\Delta) - \frac{c_\Delta}{1+c_\Delta} \right] \qquad (3.2.60)$$

where $c_\Delta = r_\Delta/r_s$ is called the concentration parameter. The NFW model can also be expressed by parameters: $(M_\Delta, c_\Delta)$. The density parameter $\rho_s$ can be expressed by concentration parameter as follows.

$$\rho_s = \frac{\Delta}{3}\rho_{\rm crit}(z_L)\frac{c_\Delta^3}{\ln(1+c_\Delta) - c_\Delta/(1+c_\Delta)} \qquad (3.2.61)$$

The convergence and the averaged convergence are given by Bartelmann [58] and Wright *et al.* [59]

$$\kappa(\theta) = \kappa_s f\left(\frac{\theta}{\theta_s}\right), \qquad (3.2.62)$$

$$\bar{\kappa}(\theta) = 2\kappa_s g\left(\frac{\theta}{\theta_s}\right)\left(\frac{\theta}{\theta_s}\right)^{-2}, \qquad (3.2.63)$$

where $\kappa_s = 2\delta_c \rho_{\rm crit} r_s \Sigma_{\rm crit}^{-1}$ and $\theta_s = r_s/D_L$. The function $f(x)$ and $g(x)$ are then given by

$$f(x) = \begin{cases} \dfrac{1}{1-x^2}\left(-1 + \dfrac{2}{\sqrt{(1-x^2)}}\operatorname{arctanh}\sqrt{\dfrac{1-x}{1+x}}\right) & x < 1 \\[3mm] \dfrac{1}{3} & x = 1, \\[3mm] \dfrac{1}{x^2-1}\left(1 - \dfrac{2}{\sqrt{(x^2-1)}}\arctan\sqrt{\dfrac{x-1}{x+1}}\right) & x > 1 \end{cases} \qquad (3.2.64)$$

$$g(x) = \ln\left(\frac{x}{2}\right) + \begin{cases} \dfrac{2}{\sqrt{(1-x^2)}}\operatorname{arctanh}\sqrt{\dfrac{1-x}{1+x}} & x < 1 \\[2ex] 1 & x = 1 \\[2ex] \dfrac{2}{\sqrt{(x^2-1)}}\arctan\sqrt{\dfrac{x-1}{x+1}} & x > 1 \end{cases} \qquad (3.2.65)$$

## Non-circular lenses

There are no exactly circular lenses in nature except a star, which we do not treat here. Any lensing object is somehow non-circular. Usually rather than giving the explicit distribution for the surface mass density or 3-dimensional mass density, the gravitational potential is given. Several interesting models exist for non-circular potentials. As far as an elliptical potential is concerned, the general model contains at least five parameters (an indicator of the strength of the lens such as the Einstein angle in the circular case, core radius, ellipticity, the power law index of density falloff and the angle of the major axis from a fixed direction). For example, the Tilted Plummer Elliptical potential is given by

$$\psi(x, y) = \frac{\alpha_E^2}{\eta}\left(\frac{w_p^2 + r_{ep}}{\alpha_E^2}\right)^{\eta/2} \qquad (3.2.66)$$

where $r_{ep}^2 = x^2(1 - \epsilon_p) + y^2(1 + \epsilon_p)$, $w_p$ is related to the core radius, $\epsilon_p$ is the ellipticity and $\eta$ is the power law index. $\eta = 1$ corresponds to the isothermal case. The reader interested in this model should consult the paper by Grogin and Narayan [60].

Although one needs a specific potential for model fitting, general properties of elliptical models are similar for any known model. Figure 3.9 shows small circular sources in the source plane and the corresponding images in the image plane. In the elliptical case, there exist inner caustic and inner critical curves. The inner caustics have points where the derivative along the curve is not well defined.

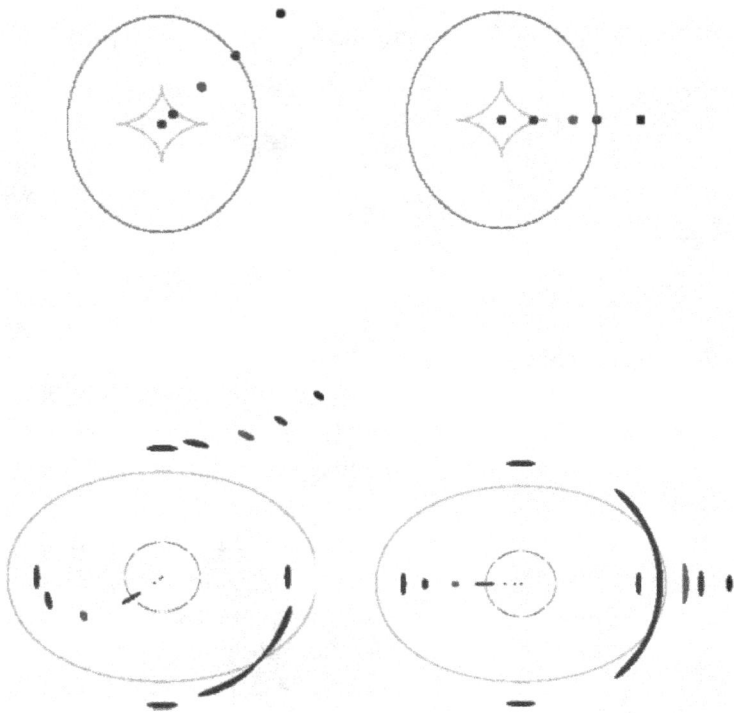

Fig. 3.9. The caustics and critical curves of an elliptical lens with sufficiently small core.

Such a point is called a cusp. Five images exist when the source is inside the inner caustic. Three images exist when the source is between the inner and outer caustics. As seen in Fig. 3.9, when a sufficiently small finite source approaches the cusp from the inside, three images (two images parallel along and outside the caustics, one perpendicular to and inside the caustics) approach a corresponding point on the critical curve and vanish when the source passes through the cusp. When the source approaches a cusp from the outside, the opposite phenomenon occurs and three images seem to appear from the corresponding point on the inner critical curve. See Blandford and Kochanek [61] for a more systematic understanding of caustic and critical curves.

## 3.3. Strong Lensing

### 3.3.1. *Methods of solving the lens equation: LTM and non-LTM*

In strong gravitational lensing, there are a number of observables such as the relative positions cf images, relative fluxes of images and time delays of luminosity changes between lensed images. These observables depend not only on the mass distribution of the source but also on the cosmological parameters. There are essentially two methods for constructing mass models of the source. The first is called the parametric method. which assumes a simple but physically reasonable model of potential such as an isothermal sphere, NFW model, elliptical pseudo-isothermal model and so on with parameters that have clear physical meanings. Physically reasonable means that one infers the shape of the lensing potential from the image position as well as the lens position and shape assuming the existence of a strong correlation between light and mass. Thus, this method is called the light traces mass (LTM) method. In the case of a galaxy lens, the relative position of the images and the existence of arcs gives us enough information to determine the form of the potential.

The other method is called the non-parametric method, which is also referred to as the "grid-based" method or non-LTM method because one does not assume the potential shape from the beginning. This method has become possible and popular since deep observations using 8–10 m telescopes and the HST reveal detailed shape information of images for galaxy lenses, and a multitude of arcs and images for cluster lenses. Such a wealth of information allows us to establish the grid by grid correspondence between the image plane and the source plane. The basic principle of this method is the conservation of the surface brightness. Namely, the surface brightness of two grids related by the lens equation is the same.

We will not go into further details of these two methods, partly because there are many strong lens model software codes using both

LTM and non-LTM methods, which are publicly available. Instead we shall focus on the basic properties of strong gravitational lensing. For more detailed treatment of strong lensing, see an extensive review by Kochanek [63].

### 3.3.2.  *Image magnification*

In strong lensing with multiple images, the ratio of the luminosity between images is an important observable and is used to constrain the mass distribution on the local scale inside the lensing galaxy.

To derive the flux ratio between images, it should be noted that gravitational lensing does not change the surface brightness. Therefore, the magnification of the image compared with the source is just the ratio of the size between the image and source, and thus the magnification of the image is the inverse of the determinant of the Jacobian. We also define the magnification matrix as the inverse matrix of the Jacobian

$$M = A^{-1} \qquad (3.3.1)$$

The magnification of an image is not known because we do not know the luminosity of the source. However, since there are multiple images, the ratio of magnification is observable and is an important constraint of the mass model of the lensing object. Suppose we have two images A and B, then using the magnification matrix, we have

$$\delta\boldsymbol{\theta}_B = M(\boldsymbol{\theta}_B)_{ij}\delta\beta = M(\boldsymbol{\theta}_B)M^{-1}(\boldsymbol{\theta}_A)\delta\boldsymbol{\theta}_A \equiv M_{BA}\delta\boldsymbol{\theta}_A. \qquad (3.3.2)$$

The flux ratio is very useful to constrain the lens model. Although a simple smooth lens potential is enough to predict image positions consistent with observation, this is not the case for the flux ratio between images. This may be regarded as an indication of the existence of dark matter substructure in the halo of the lensing galaxy [65, 66].

### 3.3.3. *Time delays*

Another possible observable in strong lensing is time difference (or time delay) between arrival of light rays in multiple images. (Refsdal [9]). Since the photons from the source travel along different paths to make the different images, the time taken by the photon from the source to the observer differs from image to image, resulting in an observable time delay. Time delays have been observed for a number of systems, and were used as a method to measure the global Hubble parameter before WMAP measured the Hubble parameter to within several percent accuracy. The method is free from any empirical assumptions associated with the method of the distance ladder, but contains an indeterminacy inherent in gravitational lensing explained below. Nonetheless, this method is potentially very useful if we have enough information to determine the lens mass distribution accurately enough.

Accurate determination of Hubble parameter attracted much attention recently because significant difference have been reported between the Hubble parameter determined using CMB observation and the m-z relation of type Ia supernovae. The reported difference is called Hubble tension [199]. In order to resolve or confirm this tension, an independent method for measuring Hubble parameter such as gravitational lensing is required.

We therefore explain in detail the method of measuring Hubble parameter using the time-delay measurement, and then report recent observations. If a source produces a short burst of light, then the photons of the burst will arrive at a time

$$\tau = \frac{1 + z_L}{H_0} \frac{d_L d_S}{d_{LS}} \left[ \frac{1}{2}(\theta - \beta)^2 - \psi(\theta) \right] \qquad (3.3.3)$$

where $d_X$ is the angular diameter distance in unit of $c/H_0$. The first term is just the result of the increase in the path length and the second term is the result of a decrease in the speed of light near the gravitating object.

The time delay between the two images A and B is given by

$$\Delta_{AB} = \tau_A - \tau_B$$
$$= \frac{1 + z_L}{H_0} \frac{d_L d_S}{d_{LS}}$$
$$\left[ \frac{1}{2} \left( (\theta_A - \beta)^2 - (\theta_B - \beta)^2 \right) - (\psi(\theta_A) - \psi(\theta_B)) \right] \quad (3.3.4)$$

For practical application of time delays, we have to take into account the fact that the lens galaxy is sometimes a member of a small group of galaxies or is under the influence of nearby galaxies. In that case, we consider these effects by the external potential in the following form:

$$\psi_c(\boldsymbol{\theta}) \simeq \psi_c(\mathbf{0}) + \nabla \psi_c(\mathbf{0}) \cdot \boldsymbol{\theta} + \frac{1}{2} \kappa_c(\mathbf{0}) \left( \theta_x^2 + \theta_y^2 \right)$$
$$+ \frac{1}{2} \gamma_{1c}(\mathbf{0}) \left( \theta_x^2 - \theta_y^2 \right) + \gamma_{2c}(\mathbf{0}) \theta_x \theta_y \quad (3.3.5)$$

where we take the origin at the position of the lens galaxy and assume that the length scale of the external field is large compared with the separation of two images and thus neglect the higher order terms. The coefficients are given by

$$\kappa_c = \frac{1}{2} (\psi_{c,xx} + \psi_{c,yy}), \quad \gamma_{1c} = \frac{1}{2} (\psi_{c,xx} - \psi_{c,yy}), \quad \gamma_{2c} = \psi_{c,xy}$$
$$(3.3.6)$$

It is easy to show that the first two terms of the above expansion does not change any observables in the strong lensing system. The third term expresses the contribution from the uniform sheet. It does have any effect on the time delay.

We now show that a uniform sheet will bring an ambiguity into the Hubble constant determination. We consider the transformation of the potential

$$\psi \to (1-c)\psi + \frac{c}{2}(\theta_x^2 + \theta_y^2) \tag{3.3.7}$$

with a constant $c$. This changes the bending angle as follows:

$$\alpha \to (1-c)\alpha + c\theta. \tag{3.3.8}$$

The lensing equation tells us that $\theta$ does not change by changing the source position $\beta \to (1-c)\beta$. This will change the magnification matrix as $M \to M/(1-c)$, but the relative magnification matrix $M_{BA} = M_B M_A^{-1}$ does not change. However, the time delay equation (62) tells that the combination $H_0 \Delta \tau_{BA}$ changes as above

$$H_0 \Delta \tau_{BA} \to (1-c)H_0 \Delta \tau_{BA} \tag{3.3.9}$$

This means that the observed time delay does not change if we can change the distance scale at the same time as

$$H_0 \to (1-c)H_0 \tag{3.3.10}$$

Thus even if we determine the lens model by image positions, as well as the ratio of relative magnification between images and time delay, there remains an ambiguity in the measurement of Hubble parameter

$$H_0 = 100h(1 - \kappa_c(0)) \quad [\text{km/s/Mpc}] \tag{3.3.11}$$

The ambiguity $\kappa_c(0)$ is the projected surface mass density near the lens position. Weak lensing analysis in the region around the lens galaxy is a direct method to determine $\kappa_c$. In fact there have been attempts to measure $\kappa_c(0)$ in QSO0957+561 A, B by weak lensing [73, 74].

It has been shown that a single gravitational lens with well-measured time delays and an accurate mass model of the lensed object including surrounding area can be used to measure time delay distances defined as $D_\Delta \equiv (1 + z_L)D_L D_S/D_{LS}$ to 6–7% total uncertainty (random and systematic Suyu *et al.* [68]). In fact, Suyu *et al.* [69] have shown that in quadruple lens systems the lens model inaccuracy can be less important than the inaccuracy in the time delay measurements. Their result is based on detailed modeling of the gravitational lens RXJ1131-1231 with the spatially extended Einstein ring observed by the HST and well-measured time delays between its multiple images. Wong *et al.* [217] analyzed six multi-image gravitational lens systems of quasars with measured time delays and Fig. 3.10 shows the joint probability distribution of the Hubble parameter. The estimated value is $H_0 = 73.3^{+1.7}_{1.8}\,\mathrm{km\,s^{-1}Mpc^{-1}}$ with an accuracy of 2.4%. This agrees with the value of the Hubble parameter measured using the m-z relation of Type Ia supernovae, but it contradicts the value of the Hubble parameter estimated from the detailed observation of the CMB mentioned above by $3.1\sigma$.

On the order of 100 new quadruple lensed quasars are expected to be discovered in the near future with imaging surveys such as the Subaru Hyper Supreme-Cam Survey, Pan-STARRS-1 [70], the Dark Energy Survey (DES) [71]. When their time delay data are available, it is expected that together with supernovae and cosmic microwave background information, we can improve the dark energy figure of merit by almost a factor of 5, and determine the matter density parameter $\Omega_{m,0}$ to 0.004, the Hubble parameter $H_0$ to 0.7%, and the dark energy equation of state time variation parameter $w_a$ to $\pm 0.26$, systematics permitting (Linder [72]).

### 3.3.4. *Strong lens model of JWST galaxy cluster SMACS J0723.3-7327*

The James Webb Space Telescope (JWST) was launched in 25th December 2021 more than 30 years after Hubble Space Telescope (HST), and a month later reached the halo orbit around the lagrange point $L2$ 1.5 million km from Earth. The observable wavelength is

$H_0 \in [0, 150] \quad \Omega_{\mathrm{m}} \in [0.05, 0.5]$

$H_0 : 71.0^{+2.9}_{-3.3}$
$H_0 : 78.2^{+3.4}_{-3.4}$
$H_0 : 71.7^{+4.8}_{-4.5}$
$H_0 : 68.9^{+5.4}_{-5.1}$
$H_0 : 71.6^{+3.8}_{-4.9}$
$H_0 : 81.1^{+8.0}_{-7.1}$

$H_0 : 73.3^{+1.7}_{-1.8}$

- All
- B1608 (Suyu+2010, Jee+2019)
- RXJ1131 (Suyu+2014, Chen+2019)
- HE0435 (Wong+2017, Chen+2019)
- J1206 (Birrer+2019)
- WFI2033 (Rusu+2019)
- PG1115 (Chen+2019)

probability density

$H_0 \ [\mathrm{km\,s^{-1}\,Mpc^{-1}}]$

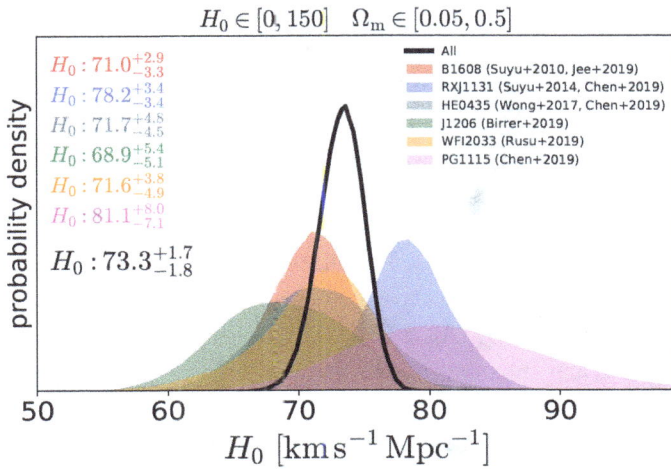

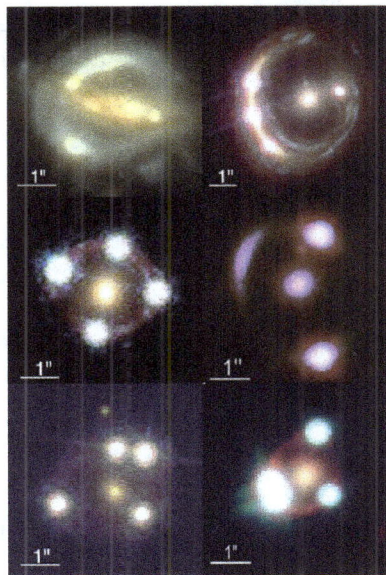

Wong et al 2022

Fig. 3.10. The joint probability distribution of the Hubble parameter by the observation of time delay in 6 multi-image gravitational lens systems.

in the infrared region of $0.6-2.8\,\mu m$ and provides unprecedented observations of high-redshift galaxies in terms of both sensitivity and angular resolution as expected. As part of the Early Release Observations, the JWST team publicly released the observations of its first cosmic targets, including the galaxy cluster SMACS J0723.3–7327.

The strong lens model is constructed by Caminha *et al.* [3] where they have employed the software lenstool [19, 32, 33] to model the mass distribution of SMACS J0723. The lenstool is one of the so-called Light trace mass software. It was first described in 1993 and has been continually upgraded. Lenstool consistes of an elliptical cluster-scale dark matter halo (six free parameters) as a fiducial halo mass model and a truncated spherical isothermal mass profile for each cluster galaxy member (two free parameters with a constant $M/L$ scaling relation for the members) and an external shear (two free parameters). The model constraints are the positions of multiple images and (when available) spectroscopic redshifts. For families of multiple images with no spectroscopic informations they are used variables to improve the fitting.

JWST increases the number of lensed images. In fact, 30 new multiple images from 11 background galaxies are identified. These images are very faint in the optical wavelength and no clear HST counterpart. Although these new families of images do not have spectroscopic redshifts, they increases by a factor of two the number of model constraints compared to the previous lens model. Fig. 3.11 shows the new lens images detected by JWST and the critical line for a source at $z = 10$ overlaid on the JWST/NIRCam imaging.

This is just one example indicating the power of JWST, and JWST will reveal more precise mass distributions of many galaxy clusters in the future, which will provide more detailed information on structure formation, such as the nature of dark matter and its relationship with baryonic matter.

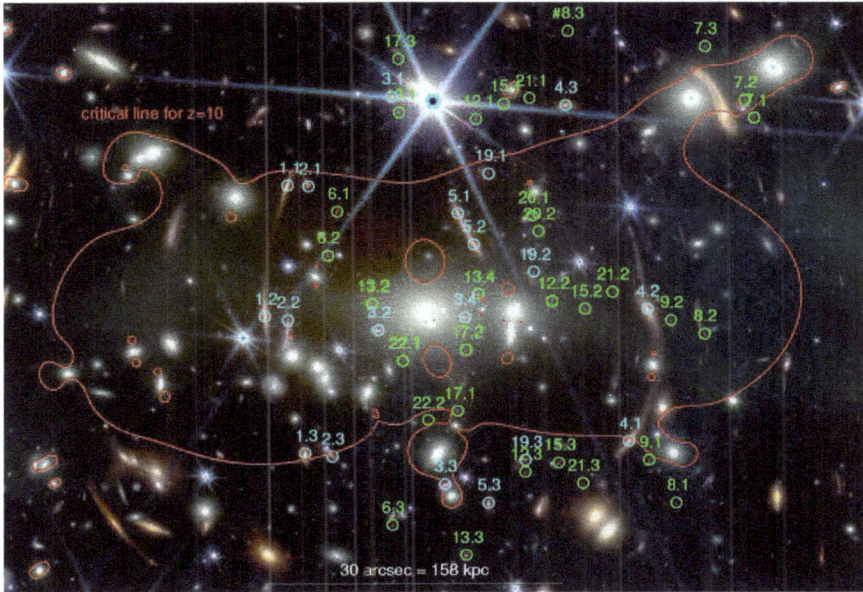

Fig. 3.11. JWST image of a cluster SMACS J0723 ($z = 0.29$) and critical curves for the source redshifr $z = 10$.

### 3.3.5. *Lens statistics*

The statistics of strong lensing events have interesting applications to cosmology. There have been many studies of statistics using quasars, giant luminous arcs and recently GRAMORs. It has attracted much attention because the statistics are sensitive to the cosmological constant via the cosmological volume element (e.g. Turner [83]; Fukugita, Futamase and Kasai [37]). As shown below, a detailed and accurate knowledge of the properties of the source and lensing object is necessary to use lens statistics as a useful method to measure the cosmological parameters and the redshift evolution of the lensing galaxies (e.g. Kochanek [84]; Ofek *et al.* [85]; Chae and Mao [86]; Matsumoto and Futamase [37]; Chae [88]; Cao and Zhu [101]). Furthermore, it has been shown that sample incompleteness can bias the results significantly (Capelo and Natarajan [89]). Recent large scale galaxy surveys such as SDSS overcame these difficulties and made the statistics a useful tool in observational cosmology (see Oguri *et al.* [90]).

Here, we explain the basic formulation of quasar statistics as an example of lens statistics. Consider the probability that a quasar with redshift $z_s$ is lensed by a lensing object with redshift range $z_L \sim z_L + dz_L$. Writing the cross-section as $\sigma_L$, the differential probability may be written as follows. We have

$$d\tau(z_L) = n(z_L)\sigma_L \left|\frac{cdt}{dz_L}\right| dz_L \tag{3.3.12}$$

where $n(z_L)$ is the number density of lensing objects. The cross-section is given by solving the lens equation once we choose an appropriate lens potential, and depends on the redshifts of the source and the lens as well as the relative position of the source and lens.

$$\sigma_L = \int_S d^2\beta \frac{\Phi\left(\frac{L}{\mu}\right)}{\mu\Phi(L)} \tag{3.3.13}$$

where the integration region $S$ in the source plane is over the position where multiple images are produced. Here, we include the magnification bias. This is necessary because some observed lensed quasars are magnified and become brighter than a limited magnitude. Thus, we need a magnification factor $\mu$ for each source position and the quasar luminosity function $\Phi(L)$. The cosmological dependence is in the quantity $cdt/dz_L$, which is calculated in FRW geometry to be

$$\frac{cdt}{dz_L} = \frac{c}{H_0(1+z_L)} \frac{1}{\sqrt{\Omega_{m,0}(1+z_L)^3 + (1-\Omega_{m,0}-\Omega_{\Lambda,0})(1+z)^2 + \Omega_{\Lambda,0}}} \tag{3.3.14}$$

where $\Omega_{m,0}$ is the matter density parameter and $\Omega_{\Lambda,0} = \Lambda/3H_0^2$ is the normalized cosmological constant.

In practical applications, the number density of the lens object is calculated using the luminosity function $\Phi(L) = dn(L)/L$ or the velocity function $\psi(\sigma_v) = dn(\sigma_v)/d\sigma_v$. For example, using the velocity function the differential probability of lensing per unit redshift with an image separation between $\theta$ and $\theta + d\Delta\theta$ is given by

$$\frac{d^2\tau}{d\Delta\theta dz_L} = \int_0^{z_S} d\sigma_v \phi(\sigma_v)(1 + z_L)^3 \frac{cdt}{dz_L} \frac{d\sigma_L}{d\Delta\theta} \qquad (3.3.15)$$

A recent study based on the lens sample from the Sloan Digital Sky Survey Quasar Lens Search (SQLS) (Oguri *et al.* [90]) used 19 lensed quasars selected from 50,836 source quasars where the velocity dispersion of the lensing galaxies suggested by SDSS data (Bernardi *et al.* [91]) and the quasar luminosity function for $0.4 < z < 2.3$ measured by the combined analysis of SDSS and 2dF data (Croom *et al.* [92]) are used. It is shown that the SQLS sample constrains the cosmological constant to $\Omega_{\Lambda,0} = 0.79^{+0.06}_{0.07}$ (stat.)$^{+0.06}_{0.06}$ (syst.) for a flat universe. It is also shown that the dark energy equation of state parameter is consistent with $w = 1$ when the SQLS data is combined with constraints from baryon acoustic oscillation (BAO) measurements or results from the Wilkinson Microwave Anisotropy Probe (WMAP). Furthermore, no redshift evolution of the galaxy velocity function at $z \leq 1$ is found.

## 3.4. Weak Lensing

Although strong lensing is an important and spectacular phenomenon, it is a rare event and most light rays from distant sources propagate through relatively weak gravitational fields. In this case, the intervening gravitational fields only slightly distort the shape of background sources, resulting in a systematic distortion pattern of background source images known as weak gravitational lensing. In the past decade, weak lensing has become a powerful, reliable method to measure the distribution of matter in clusters, dominated by invisible dark matter (DM), without assumptions concerning the physical and dynamical state of the system. Not only clusters, but also weak lensing due to large scale structure, attracts much attention because it carries information regarding the growth of structure which is affected by the nature of dark energy. For a general review of weak gravitational lensing, see Bartelmann and Schneider [111] and Umetsu [112].

### 3.4.1. *Basic method*

We shall explain one method of mass reconstruction using weak lensing mass observations in some detail. For clarity of the argument we treat the simplest situation in this subsection. The complexities are mentioned later.

First, we select a sample of background galaxies for each lensing object considered. Since the lensing signal depends on the redshifts of the sources, it is necessary to have redshift information for accurate determination of the mass distribution of the lensing object. For example, in a weak lensing analysis of galaxy clusters, the contamination by faint member galaxies dilutes the lensing signal, in particular in cluster central regions [113]. In the case of cluster lensing, the background galaxies are selected by color information. Namely, their colors are redder than the color-magnitude sequence of cluster member galaxies due to large k-corrections (for more detail, see Medezinski *et al.* [114], Okabe and Umetsu [42]. However, more detailed redshift information regarding the background galaxies is necessary for accurate determination of cosmic shear (see below).

Once we define the sample of background objects, the next step is to measure the shape of each background object. Several shape measurement schemes have been developed. We concentrate here on the so-called moment method, which uses multiple moments of brightness distribution to characterize the shape of the background objects (see KSB [41] and its improved version [93–96]. There are other methods of shape measurements not using multiple moments such as the methods based on the modeling of intrinsic galaxy profile predicted by astrophysically motivated galaxy models [97, 98]. For the detail in modern shape measurement methods, we refer the interested reader to some review such as Kitching [99] and Mandelbaum [100]. and references therein.

In the moment method, we measure the quadrupole moments of surface brightness for each background object

$$Q_{ij}^{(\text{obs})} = \int d^2\theta \Delta\theta_i \Delta\theta_j I(\boldsymbol{\theta}) \tag{3.4.1}$$

where $I(\boldsymbol{\theta})$ is the surface brightness distribution and $\Delta\theta_i = \theta_i - \theta_{0,i}$ with $\theta_{0,i}$ the center of the image. Below, we define the center as the point at which the dipole moment vanishes. Then we define a purely spin-2 quantity from the measured quadrupole moments

$$\epsilon^{(\text{obs})} = \left( \frac{Q_{11}^{(\text{obs})} - Q_{22}^{(\text{obs})}}{Q_{11}^{(\text{obs})} + Q_{22}^{(\text{obs})}}, \frac{2Q_{12}^{(\text{obs})}}{Q_{11}^{(\text{obs})} + Q_{22}^{(\text{obs})}} \right) \tag{3.4.2}$$

which is called the ellipticity. For another definition of ellipticity, see Bartelmann and Schneider [101]. Similarly, we define the quadrupole moments $Q_{ij}^{(\text{s})}$ and ellipticity $\epsilon^{(\text{s})}$ for each of the corresponding sources which are not observable. Using the lens equation we find the relationship between the ellipticity of the observed image and of the intrinsic source

$$\epsilon^{(\text{s})} = \frac{\epsilon^{(\text{obs})} - 2g + g^2 \epsilon^{(\text{obs})}}{1 + |g|^2 - 2\text{Re}(g\epsilon^{(\text{obs})}{}^{*})} \tag{3.4.3}$$

In the weak lensing limit

$$\epsilon^{(\text{s})} \simeq \epsilon^{(\text{obs})} - 2g \tag{3.4.4}$$

Since we can expect that the intrinsic ellipticities of nearby galaxies on the sky do not correlate with each other, we can average over a certain number of background galaxies to obtain

$$\langle \epsilon^{\text{obs}} \rangle \simeq 2g + O\left( \frac{\sigma_e}{\sqrt{N}} \right) \tag{3.4.5}$$

where $\sigma_e \simeq 0.3 - 0.4$ is the dispersion of the intrinsic ellipticity of galaxies, and $N$ is the number of background galaxies averaged over. The more galaxies we average over, the more accurate the estimation of the gravitational shear we have. In the best seeing condition at the summit of Mauna Kea we may have $N \geq 50$ per arcmin$^2$. In a realistic situation, the number is less than 30 or so. In the case of cluster lensing, the lensing signal is on the order of 0.1 and thus we can safely detect the lensing signal by clusters.

Once we have the averaged shear field we can convert it to a convergence map which is the normalized surface mass density using the relationship between the convergence and the shear. Although this is the essence of weak lensing mass reconstruction, things are much more complicated for real observations.

## Shape measurements

In practical observations of weak lensing, the observed image is not the lensed image but rather the image smeared by various sources of noise such as the distortion by atmospheric turbulence, imperfect optics, pixelization by the CCD chip, and photon noise. Thus, we have to correct these effects before applying a mass reconstruction method.

It is supposed that these effects are expressed by the point spread function (PSF) $P(\boldsymbol{\theta})$ as follows:

$$I^{(\mathrm{obs})}(\boldsymbol{\theta}) = \int d^2\psi I(\boldsymbol{\psi})P(\boldsymbol{\theta} - \boldsymbol{\psi}) \qquad (3.4.6)$$

In order to improve the accuracy of weak lensing analysis, we need to reduce various sources of noise and to correct the PSF effects without introducing other errors.

In this review, we focus our attention on PSF correction in some detail because this has the most important effect in weak lensing analysis. There are several methods for PSF correction available today. Here, we employ the moment-based approach (see Kaiser *et al.* [41] hereafter KSB). In realistic observations, since the observed shapes of background galaxies are rather noisy, particularly for small faint galaxies, we need to introduce a weight function $W$ to reduce the random noise in the outer part of the image, so that

$$Q_{ij}^{(\text{obs})} = \int d^2\theta\, \theta_i\theta_j I^{(\text{obs})}(\boldsymbol{\theta}) W\left(\frac{\boldsymbol{\theta}}{\sigma^2}\right) \tag{3.4.7}$$

where $\sigma$ is a typical scale for the galaxy. The simple choice is the spherical Gaussian shape:

$$W\left(\frac{\boldsymbol{\theta}}{\sigma}\right) \propto e^{-\frac{|\boldsymbol{\theta}|^2}{\sigma^2}} \tag{3.4.8}$$

It is known that this choice introduces a systematic error in measuring the shape that overestimates the shear for images with large ellipticities. The error is overcome by introducing an elliptical window function (see Okura and Futamase [96]).

In KSB method and its modified versions it is assumed that PSF is expressed as a convolution of a kernal $q$ with a small anisotropy and an isotropic function $P^{(\text{iso})}$ as follows.

$$P(\boldsymbol{\theta}) = \int d^2\theta'\, q(\boldsymbol{\theta'}) P^{(\text{iso})}(\boldsymbol{\theta} - \boldsymbol{\theta'}) \tag{3.4.9}$$

where $q, P^{(\text{iso})}$ are normalized and have vanishing 1-order moments.

$$\int d\theta\, q(\boldsymbol{\theta}) = \int d\theta\, P^{(\text{iso})}(\boldsymbol{\theta}) = 1 \tag{3.4.10}$$

$$\int d\theta\theta_i q(\boldsymbol{\theta}) = \int d\theta\theta_i P^{(\text{iso})}(\boldsymbol{\theta}) = 0 \tag{3.4.11}$$

(i) Anisotropic PSF correction

We define

$$I^{(\text{iso})}(\boldsymbol{\theta}) = \int d^2\theta'\, I(\boldsymbol{\theta'}) P^{(\text{iso})}(\boldsymbol{\theta} - \boldsymbol{\theta'}) \tag{3.4.12}$$

Then the observed intensity may be written as follows:

$$I^{(\mathrm{obs})}(\boldsymbol{\theta}) = \int d\theta' q(\boldsymbol{\theta} - \boldsymbol{\theta}') I^{(\mathrm{iso})}(\boldsymbol{\theta}') \qquad (3.4.13)$$

Assuming a small anisotropy, we ignore higher multipole moment except quadrupole moment of the kernel $q$.

$$q_{ij} = \int d^2\theta q(\boldsymbol{\theta}) \theta_i \theta_j \qquad (3.4.14)$$

Then the relation between the quadrupole moment $Q^{(\mathrm{iso})}$ calculated from $I^{(\mathrm{iso})}$ and the observed quadrupole moment is derived as follows.

$$Q_{ij}^{(\mathrm{iso})} = Q_{ij}^{(\mathrm{obs})} - \frac{1}{2} Z_{ijk\ell} q_{k\ell} \qquad (3.4.15)$$

where

$$Z_{ijk\ell} = \int d^2\theta I^{(\mathrm{obs})}(\boldsymbol{\theta}) \frac{\partial^2}{\partial\theta_i \partial\theta_j} \left[ \theta_k \theta_\ell W \left( \frac{|\boldsymbol{\theta}|^2}{\sigma^2} \right) \right] \qquad (3.4.16)$$

Using the definition of ellipticity and the above expression, we find the ellipticity corrected by the anisotropic PSF effect is

$$\epsilon^{(\mathrm{iso})} = \epsilon^{(\mathrm{obs})} - P_{ij}^{(\mathrm{sm})} q_j \qquad (3.4.17)$$

where we define $q_1 = q_{11} - q_{22}, q_2 = 2q_{12}$. $P^{(\mathrm{sm})}$ is called the smear polarizability which expresses the response of the ellipticity against the anisotropic PSF effect (a detailed expression may be found in the original paper by KSB and the review by Bartelmann and Schneider). Since $P^{(\mathrm{sm})} \propto r_g^{-2}$ (where $r_g$ is the Gaussian scale of the object), the PSF anisotropic effect becomes important for objects with small $S/N$.

The anisotropic kernal $q_i$ is given by applying the equation to stars. Since stars are in our galaxy and do not suffer lensing. Thus, their images have no ellipticity after isotropic smearing, $\epsilon^{(*,\mathrm{iso})} = 0$ and we have

$$q_i^* = (P^{(*,\mathrm{sm})})_{ij}^{-1} \epsilon_j^{(*,\mathrm{obs})} \qquad (3.4.18)$$

where the quantities with an asterisk denote quantities associated with stars. To obtain the smooth map of $q_i$ other than the position of stars, the co-added mosaic image is divided into small regions with scales determined by the typical coherent scale of PSF anisotropy patterns. Then, PSF anisotropy in individual region is well described by fairly low-order polynomials. Typically, the rms value of stellar ellipticities is reduced from a few % to $(4-8) \times 10^{-3}$ after anisotropic PSF correction. Figure 3.12. shows the distribution of stellar ellipticity before and after the PSF anisotropy correction using the Subaru data for the cluster Abell 1689 (Umetsu and Broadhurst [115]).

(ii) Isotropic PSF correction

Next, we have to correct for magnification due to the PSF effect. As with the anisotropic correction, the isotropic correction can be written as a response of the ellipticity against the shear $g_i$ under the assumption that the distortion is small.

$$\epsilon^{(\mathrm{obs})} = \epsilon^{(s)} + P_{ij}^{\mathrm{sm}} q_j + P_{ij}^{(\mathrm{sh})} g_j \qquad (3.4.19)$$

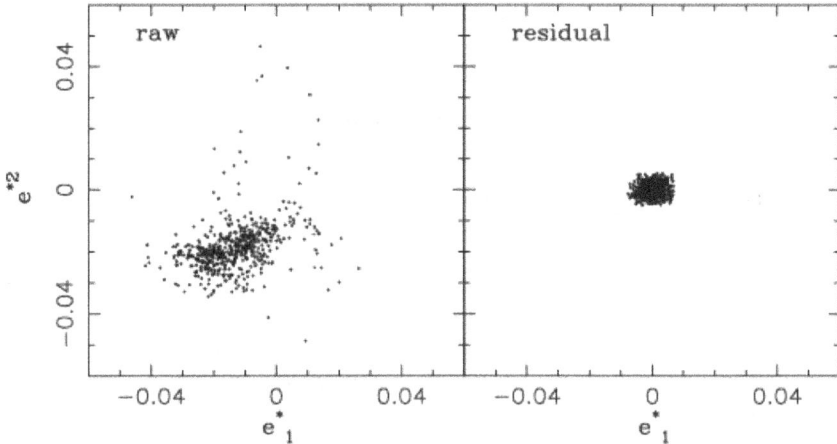

Fig. 3.12. Stellar ellipticity distributions before and after PSFanisotropy correction.

where $P^{\mathrm{sh}}$ is the responce matrix called shear polarizability. The detailed expression may be found in the original paper which describes KSB. By using star images, we have

$$q_i = -(P^{(*,\mathrm{sm})})^{-1}_{ik} P^{(*,\mathrm{sh})}_{kj} g_j \qquad (3.4.20)$$

Finally, we finally have

$$\epsilon^{(s)} = \epsilon^{(\mathrm{obs})} - P^{(\mathrm{sm})}_{ij} q_j - P^g_{ij} g_j \qquad (3.4.21)$$

where we have defined

$$P^g_{ij} = P^{(\mathrm{sh})}_{ij} - P^{(\mathrm{sm})}_{ik} (P^{*,sm})^{-1}_{k\ell} P^{(*,\mathrm{sh})}_{\ell j} \qquad (3.4.22)$$

Since the intrinsic ellipticity of background galaxies do not have a correlation, averaging gives us

$$\langle g_i \rangle = \langle (P^g)^{-1}_{ij} (\epsilon^{(\mathrm{obs})}_j - P^{(\mathrm{sm})}_{ij} q_j) \rangle \qquad (3.4.23)$$

with $q_i = (P^{(*,\mathrm{sm})})^{-1}_{ij} \epsilon^{(*,\mathrm{obs})}_j$.

As seen from the above argument, KSB has adopted some assumptions such as a small anisotropy in $q$ and the use of the observed image $I^{(\mathrm{obs})}$ in the evaluation of smear polarizability and shear polarizability. These assumptions caused systematic errors in the original KSB weak lensing analysis.

The systematic errors in the PSF correction can be evaluated by the analysis using numerical simulation using realistic galaxy images. In general the systematic bias in the shear measurement is evaluated by the following two parameters $m$ and $c$.

$$g_i^{\mathrm{obs}} = (1 + m) g_i^{\mathrm{true}} + c, \quad (i = 1, 2) \qquad (3.4.24)$$

where $g_i^{\mathrm{true}}$ are input real shear and $g_i^{\mathrm{obs}}$ are the measured shear. $m$ is called the multiplicative bias (or calibration bias) and $c$ is called additive bias(or PSF residual). $m = 2 \times 10^{-3}$ and $c = 2 \times 10^{-4}$ are required for the application to ESA Euclid space mission of cosmic shear measurement. Various methods have been proposed and tested to achieve this goal. The details of the tests can be seen in the GREAT08 and GREAT10 challenges (Bridle *et al.* [131, 132]; Kitching *et al.* [133, 135, 138]) and STEP (Heymans *et al.* [136]; Massey *et al.* [137]), and the review by Mandelbaum [100].

In the actual measurement, the averaged shear over the observed region is estimated by introducing the weight function $u_{g,I}$ such that

$$u_{g,I} = \frac{1}{\sigma_{g,I}^2 + \alpha^2} \tag{3.4.25}$$

where $I$ labels the object in the region, the constant $\alpha$ is the softening parameter and $\sigma_g$ is the rms error of the complex distortion measurement. Usually $\alpha \simeq \langle \sigma_g^2 \rangle \simeq 0.4$ is used, which is a typical value of the mean rms $\bar{\sigma}_g$ over the background sample. Then the averaged shear is given by

$$\langle g_i(\boldsymbol{\theta}) \rangle = \frac{\sum_I \omega_g(\boldsymbol{\theta} - \boldsymbol{\theta}_I) u_{g,I} g_{i,I}}{\sum_I \omega_g(\boldsymbol{\theta} - \boldsymbol{\theta}_I) u_I} \tag{3.4.26}$$

where $g_{i,I}$ is the reduced shear estimated of the $I$th galaxy at an angular position $\boldsymbol{\theta}_I$, and $\omega_g(\theta) \propto \exp(-\theta^2/\theta_g^2)$ with $\theta_g = \mathrm{FWHM}/\sqrt{4\ln 2}$ which pixelizes the distortion data into a regular grid of pixels. The error variance for the above smoothed shear is given as

$$\sigma_{\langle g \rangle}^2(\boldsymbol{\theta}) = \frac{\sum_I \omega_{g,I}^2 u_{g,I}^2 \sigma_{i,I}^2}{\left(\sum_I \omega_{g,I} u_I\right)^2} \tag{3.4.27}$$

where $\omega_{g,I} = \omega_g(\boldsymbol{\theta} - \boldsymbol{\theta}_I)$ and the relationship $\langle g_{i,I} g_{j,J} \rangle = (1/2)\sigma_{g,I}^2 \delta_{ij} \delta_{IJ}$ has been used

The smoothing scale is chosen to optimize the weak lensing detection of target mass structures, depending on both the size of the structure and the strength of noise power ($\propto \bar{\sigma}_g^2/n_g$).

Once the smoothed shear field is thus obtained, it can be converted to the convergence field using the relation $\kappa(k) = \pi D^*(k)\gamma(k)$. Actually, this inversion method has to be applied in an infinite space. For a finite space one need to use the finite-field solution of the inversion problem for the reconstruction kernel. Such method has been developed by Seitz and Schneider [122]. For a sufficiently wide field where the data field is not dominated by a positively or negatively biased density field, both methods give the almost same convergence field. This is not the case for a nearby cluster for which we have to use a finite field method.

### 3.4.2.  *E/B decomposition*

Observations and data analysis suffer from errors and noise. In particular, weak lensing analysis needs many corrections such as for the PSF distortion to isolate a physical signal of a weak tidal gravitational field. If the corrections are not perfect, they may systematically change the signal. Thus, it is critically important to characterize the effect of systematic errors on the final results.

Here we consider a method for the evaluation of systematic errors called E/B decomposition. This method is based on the fact that gravitational shear is generated by a Newtonian potential in the lowest order, and thus the gravitational field is rotation-free.

Let us first define a new 2D vector field $\mathbf{u}$ from the reconstructed convergence $\kappa$ by

$$\mathbf{u} \equiv \nabla \kappa = \begin{pmatrix} \gamma_{1,1} + \gamma_{2,2} \\ \gamma_{2,1} - \gamma_{1,2} \end{pmatrix} \tag{3.4.28}$$

Using the vector field, we define the $E$-mode $\kappa$ and $B$-mode $\kappa$ as follows:

$$\nabla^2 \kappa^E \equiv \nabla \cdot \mathbf{u}, \tag{3.4.29}$$

$$\nabla^2 \kappa^B \equiv \nabla \times \mathbf{u} = u_{2,1} - u_{1,2} \tag{3.4.30}$$

which gives both $E$-mode and $B$-mode potentials

$$\nabla^2 \phi^{E,B} = 2\kappa^{E,B} \tag{3.4.31}$$

Note that the gravitational effect is given by the Newtonian potential $\Psi$ in the lowest order so that $\phi^E = \Psi$ and $\phi^B = 0$, but in practical observations. $\phi^E \neq \Psi$ and $\phi^B \neq 0$. The complex shear may be expressed by the complex potential $\phi = \phi_E + i\phi_B$ as follows:

$$\gamma = \left[\frac{1}{2}\left(\phi^E_{,11} - \phi^E_{,22}\right) - \phi^E_{,12}\right] + i\left[\phi^E_{,12} + \frac{1}{2}\left(\phi^B_{,11} - \phi^B_{,22}\right)\right] \quad (3.4.32)$$

This expression shows that the transformation $\gamma'(\boldsymbol{\theta}) = i\gamma(\boldsymbol{\theta})$ is equivalent to an interchange of the $E$- and $B$-modes of the original maps. By noticing that the shear transforms as $\gamma' = \gamma e^{2i\phi}$ under the rotation of an angle $\phi$, this operation is a rotation of each ellipticity by $\pi/4$ with each potential being fixed. Figure 3.13 shows the E and B mode shear.

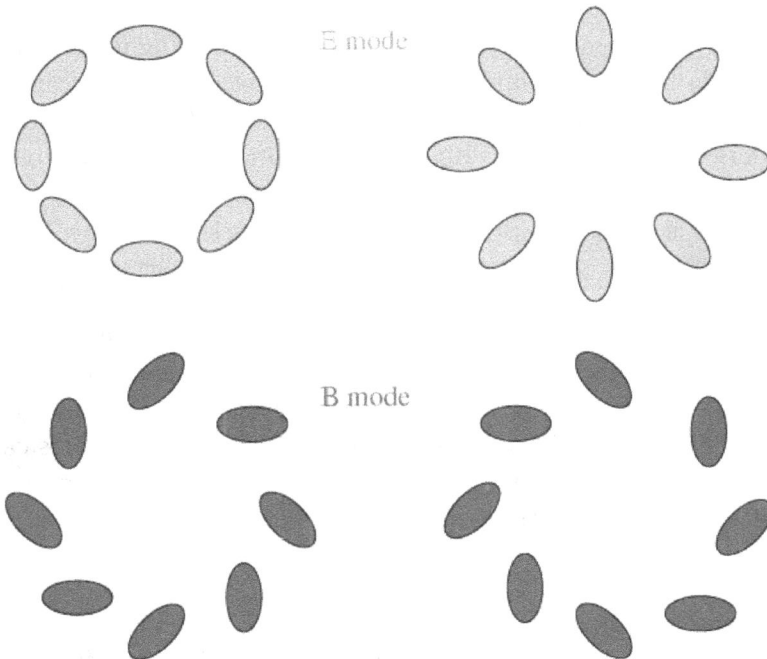

Fig. 3.13. Shear patterns of $E$- and $B$- modes.

There are several sources of $B$-mode shear. One artificial source is PSF residuals. Since the PSF correction so far proposed adopts some sort of approximation, it gives rise to $B$-mode shear. In addition to artificial $B$-modes, there are also physical $B$-modes. Intrinsic ellipticities of background galaxies contribute to the shear estimate. By assuming randomness of orientation of the intrinsic ellipticities, such uncorrelated ellipticities yield statistically identical contributions to the $E$- and $B$-modes.

By increasing the number density of background galaxies, these modes may be reduced. Another physical $B$-mode is due to the existence of intrinsic alignments, which may be generated by tidal forces between halos of galaxies. There are two kinds of intrinsic alignments. One is called II correlation. Galaxies around a large halo may have some correlation of the orientation of ellipticity by a tidal gravitational force. This effect creates both $E$- and $B$-modes. It is reasonable to assume that the effect only applies to nearby galaxies, and thus the effect is eliminated by excluding nearby galaxy pairs from the background sample [124]. Another effect is called GI correlation, which is the correlation between the intrinsic ellipticities and large scale structure [125]. This is because the ellipticites are influenced by a local density field and the same density field causes weak lensing for high-z galaxies (cosmic shear). Thus, the $B$-mode signals serve as a useful null check for systematic effects.

Finally, we highlight an ambiguity in weak lensing analyses. As pointed out above, the relationship between the convergence and the shear is not unique. Any constant mass sheet does not change the shear field which is obtained by the weak lensing limit. This is called the mass-sheet degeneracy. Actually as shown above, the observable is not shear itself but the reduced shear

$$g(\boldsymbol{\theta}) = \frac{\gamma(\boldsymbol{\theta})}{1 - \kappa(\boldsymbol{\theta})} \qquad (3.4.33)$$

This field is invariant under the following global transformation:

$$\kappa(\boldsymbol{\theta}) \to \lambda\kappa(\boldsymbol{\theta}) + 1 - \lambda, \quad \gamma(\boldsymbol{\theta}) \to \lambda\gamma(\boldsymbol{\theta}) \qquad (3.4.34)$$

where $\lambda \neq 0$ is an arbitrary constant. There is a method to break this degeneracy once one realizes that the above transformation is equivalent to the following scaling for the Jacobian matrix:

$$A(\boldsymbol{\theta}) \rightarrow \lambda A(\boldsymbol{\theta}) \qquad (3.4.35)$$

One can exploit magnification effects since the inverse of the determinant of the Jacobian is the magnification.

### 3.4.3. *Magnification bias*

Recent advances in weak lensing observations made the measurement of the magnification bias possible by measuring the depletion of the surface density of background objects. There are two effects as a result of lensing magnification. One is the expansion of the area of sky. The other is the amplification of the apparent magnitude which is given by

$$m^{\text{obs}} = m + 2.5 \log \mu \qquad (3.4.36)$$

where $m$ is the apparent magnitude without lensing. These effects change the magnitude-limited number density $n(< m_{\text{lim}})$ of the background objects, so that

$$n(< m_{\text{lim}}) = \frac{1}{\mu} n_0(< m_{\text{lim}} + 2.5 \log \mu) \qquad (3.4.37)$$

where $m_{\text{lim}}$ is the limiting magnitude and $n_0$ is the unlensed number density count per solid angler. Let us suppose that $n_0$ near the limited magnitude obeys a power law,

$$\alpha = \frac{d \log n_0(< m)}{dm} \qquad (3.4.38)$$

where $\alpha$ is the power law index around $m = m_{rm\text{lim}}$. Then we have

$$n(< m_{\text{lim}}) = n_0(< m_{\text{lim}}) \mu^{2.5\alpha - 1} \qquad (3.4.39)$$

Since $\mu > 1$, $n(< m_{\text{lim}}) < n_0(< m_{\text{lim}})$ if $\alpha < 0.4$. The ratio is

$$b = \frac{n < m_{\text{lim}})}{n_0(< m_{\text{lim}})} = \mu^{2.5\alpha-1} \simeq 1 + (5\alpha - 2) \qquad (3.4.40)$$

where we have used the weak lensing limit in the final step.

For red background galaxies at a median redshift $\sim 1$, the intrinsic power law index at faint magnitudes is observed to be relatively flat $\alpha \simeq 0.1$ and thus causes depletion of the number count. This depletion has been observed in several massive clusters (Broadhurst *et al.* [113]; Umetsu and Broadhurst [115]; Umetsu [116]). The lower figure of Fig. 3.14 shows the observed tangential shear and magnification bias of galaxy clusters by Subaru telescope. In this figure shows the radial distribution of the tangential shear (square in the upper figure) and magnification bias (circle in the lower figure) of a galaxy cluster obtained from weak gravitational lens observations with the Subaru Telescope. The magnification bias for the red background galaxy sample ($z \sim 1$) and the more distant blue background galaxy sample ($z \sim 2$).

### 3.4.4.  *Cluster mass reconstruction*

In the hierarchical structure formation in the current standard cosmological model($\Lambda$CDM), a small halo of dark matter is formed first, and then star formation progresses as baryonic matter falls into its gravitational potential. This leads to the formation of galaxies, and the gravitational merging of galactic halos to form larger structures such as galaxy groups and clusters. According to cosmological N-body numerical simulations, it is known that the average mass distribution of collisionless CDM halos created in this way can be well approximated by the NFW model [205]. Therefore, by accurately determining statistical and universal properties such as galaxy cluster counts and mass distributions from observations, we can provide important constraints on cosmology and structure formation theory. In particular galaxy clusters are the

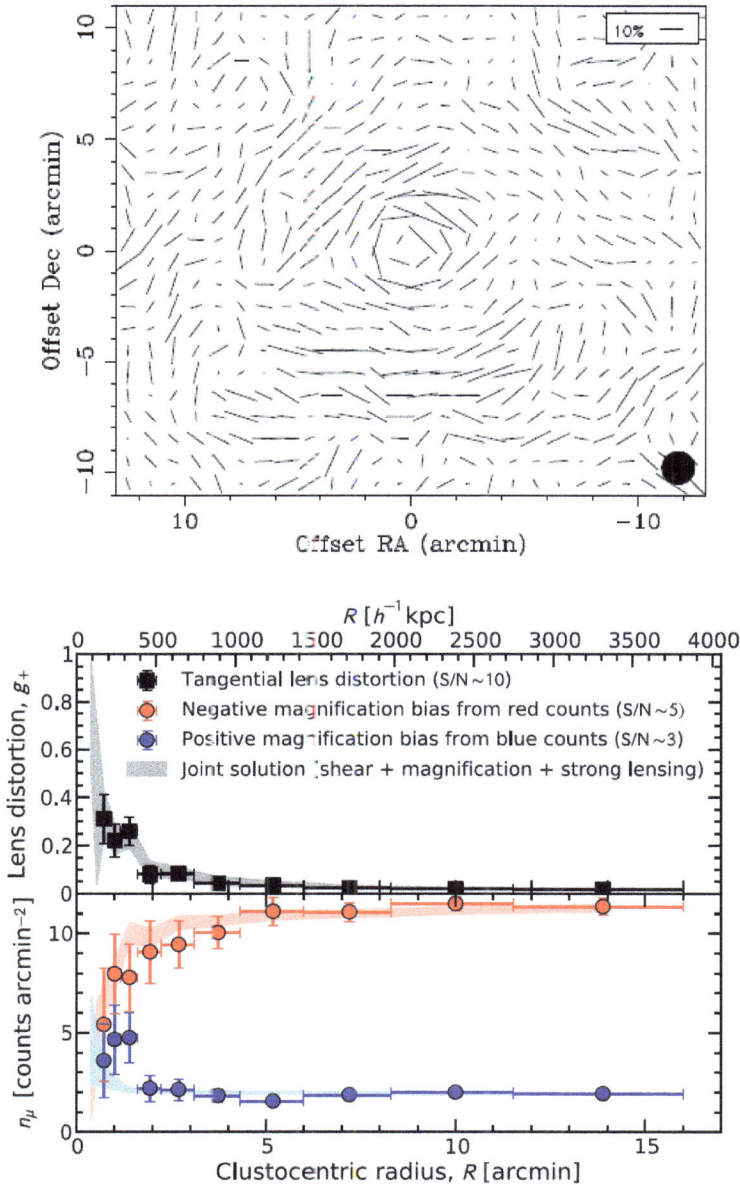

Fig. 3.14.  Observed tangential shear and magnification bias of galaxy clusters by Subaru telescope.

largest self-gravitating systems in the universe, with a total mass of more than $10^{14} M_\odot$ and most of the mass exists as dark matter, which accounts for about 80% of the entire galaxy cluster. This is the treason for the importance of the statistical study of galaxy clusters.

The most direct method for measuring the mass of galaxy clusters is gravitational lensing. Traditionally, galaxy clusters mass measurements have been performed using only weak lens effects. However, with the latest observational instruments such as HST and JWST, very deep imaging observations have detected dozens to hundreds of multiplexes per galaxy cluster in the case of massive galaxy clusters, This makes it possible to measure the precise mass distribution in the central region of individual galaxy clusters. Therefore, by using a combination of strong gravitational lensing analysis at the center and weak gravitational lensing analysis at the periphery, it is now possible to obtain an accurate mass distribution from the center to the outer edge of the cluster.

Before introducing the observation of accurate mass distribution from the center to the outer edge of galaxy clusters using both strong and weak gravitational lensing analyses, let us explain frequently used techniques for measuring the cluster mass profile in weak lensing. For a more comprehensive treatment of cluster weak lensing and its applications, see the reviews by Umetsu [215] and by Kneib and Natarajan [64].

As explained above, the tangential shear $g_+$ is a coordinate independent quantity for a given reference point. For cluster lensing, the reference point is naturally the center of the cluster, which can be determined from the symmetry of the strong lensing pattern, the X-ray centroid position, or the position of brightest cD galaxy. It is also convenient that the cross-shear $g_\times$ may be used as an estimator of the systematic error since $g_\times$ is the divergence-free, curl-type distortion pattern of background images. The reason to use tangential shear for measuring the mass profile is that the azimuthal average of the tangential shear allows us to directly compare with the theoretical model by using the relationship

$$\langle \gamma_t \rangle(\theta; \boldsymbol{\theta}_0) = \bar{\kappa}(< \theta, \boldsymbol{\theta}_0) - \langle \kappa \rangle(\theta, \boldsymbol{\theta}_0). \tag{3.4.41}$$

where $\langle \cdots \rangle$ is the azimuthal average and $\bar{\kappa}(< \theta, \boldsymbol{\theta}_0)$ is the average of the convergence within the radius $\theta$. In the following, we understand that $\theta$ is the radius from the origin $\boldsymbol{\theta}_0$ and will not write the origin. The above relationship is obtained by the definition of $\bar{\kappa}$ in the following form:

$$\theta^2 \bar{\kappa}(< \theta) = 2 \int_0^\theta d\theta' \theta' \langle \kappa(\theta) \rangle = \frac{\theta}{2\pi} \int d\psi \frac{\partial \phi}{\partial \theta} \tag{3.4.42}$$

where we have used the fact that $2\kappa = \Delta\phi$, and the 2D version of Gauss's theorem. By taking the derivative with respect to $\theta$ and using the relationship

$$\frac{\partial^2 \phi}{\partial \theta^2} = \kappa - \gamma_t \tag{3.4.43}$$

we have

$$\frac{d\theta^2 \bar{\kappa}(< \theta)}{d\theta} = \theta \bar{\kappa}(< \theta) + \theta[\langle \kappa \rangle(\theta) - \langle \gamma_t \rangle(\theta)] \tag{3.4.44}$$

This is the required relationship by noting $d(\theta^2 \bar{\kappa}/)d\theta = 2\langle \kappa \rangle$.

It is also useful to define the following quantity, over the annulus bounded by $\theta$ and $\theta_{\text{out}}$ located outside the mass to be measured:

$$\zeta(\theta, \theta_{\text{out}}) \equiv \bar{\kappa}(< \theta) - \bar{\kappa}(\theta < \theta < \theta_{\text{out}}) \tag{3.4.45}$$

where $\theta$ and $\theta_{\text{out}}$ are the inner and outer radii of the annulus centered on some position $\boldsymbol{\theta}_0$ usually taken at the position of the brightest galaxy in the cluster. This is referred to as $\zeta$-statistics (Fahlman *et al.* [155]). Using the above equations, we can derive

$$\zeta(\theta, \theta_{\text{out}}) = \frac{2}{1 - \frac{\theta^2}{\theta_{\text{out}}^2}} \int_\theta^{\theta_{\text{out}}} d\ln\theta \langle \gamma_t(\theta) \rangle. \tag{3.4.46}$$

Thus, $\zeta$ is directly obtained by observing the image distortion of background galaxies in the weak lensing regime where $|\kappa|, |\gamma| \ll 1$. Since $\bar{\kappa}(\theta < \theta < \theta_{\rm out})$ is positive definite, $\bar{\kappa}(< \theta) \geq \zeta(\theta, \theta_{\rm out})$. Thus, we can define the lowest projected mass contained within the radius $\theta$ as

$$M_\zeta(< \theta) = \pi(D_d\theta)^2 \Sigma_{\rm cr}\zeta(\theta, \theta_{\rm out}) \qquad (3.4.47)$$

By fixing the outer radius $\theta_{\rm out}$ sufficiently large (but not so large to contain neighboring clusters), $\zeta(\theta, \theta_{\rm out})$ as a function of $\theta$ is called the radial profile. This profile does not depend upon the invariance transformation. We can fit the observed radial profile with some theoretical models.

Using the technique explained above one can observe the mass distribution of galaxy clusters. Galaxy clusters are the most massive gravitationally bound systems in the universe, and their formation strongly depends on cosmological parameters and the properties of dark matter. Therefore, accurate measurements of the mass and mass distribution of galaxy clusters are essential for applying galaxy clusters to cosmological research. There are several projects aiming at an accurate measurement of the cluster mass profile. The largest samples of clusters for which weak-lensing observations are available are currently drawn from large-scale X-ray surveys and number of order 50 clusters. These surveys are the Local Cluster Substructure Survey (LoCuSS) [42, 126], the Canadian Cluster Comparison Project (CCCP) [127, 128] and the Weighing the Giants programme (WtG) [129, 130].

There have been many works on cluster lensing and related works, Here, we only mention the following three projects carried out by the Subaru telescope. We apologize for not being able to touch on many important works on weak lensing and strongly recommend reader to refer to the reviews listed above for more extensive research.

## LoCuSS

The Local Cluster Substructure Survey (LoCuSS) is aimed at measuring the mass and structure in our nearby universe as accurately as possible. by studying intensely 50 of the most massive galaxy clusters in the redshift range $0.15 < z < 0.3$ with various kind of observations. These are selected from the ROSAT All Sky Survey catalogs (Ebeling *et al.* [162, 163]; Boehringer *et al.* [164]) that satisfy $L_X[0.1 - 2.4\,\text{keV}]/E(z)^{2.7} \geq 4.2 \times 10^{44}\,\text{ergs}^{-1}, 0.15 \leq z \leq 0.30, n_H < 7 \times 10^{20}\,\text{cm}^{-2}$, and $-25° < \delta < +65°$, where $E(z) = H(z)/H_0$ is the normalized Hubble parameter. The mass maps of these clusters were obtained by weak lensing using Subaruu telescope. Here, we show the result of our measurement of the mean density profile by stacking the weak lensing signals of all clusters (Okabe *et al.* [42]).

As mentioned above, the selection of background galaxies is important for an accurate mass reconstruction. The sample of background galaxies basically consists of red galaxies with $(V - i')$ colors redder than the red sequence of cluster members. This results in just 1% contamination of the sample by foreground and cluster members.

Each individual cluster is detected at a typical peak signal-to-noise ratio $(S/N) \simeq 4$ using the 2D Kaiser–Squires mass reconstructions. The signal is stacked by the shear catalogs in physical length units centered on the respective brightest cluster galaxies (BCGs) and the average cluster mass distribution for the full sample, with a peak $S/N$ of 28. The stacked lensing mass measurement is less sensitive to cluster internal structures and 3-dimensional halo orientation. Thus, it enables us to measure an averaged cluster mass profile for the sample. When the signals from 50 clusters are combined, the number density of background galaxies becomes $266.3\,\text{arcmin}^{-2}$. The $B$-mode signal is smaller by at least an order of magnitude than the $E$-mode signal. Figure 3.15(a)

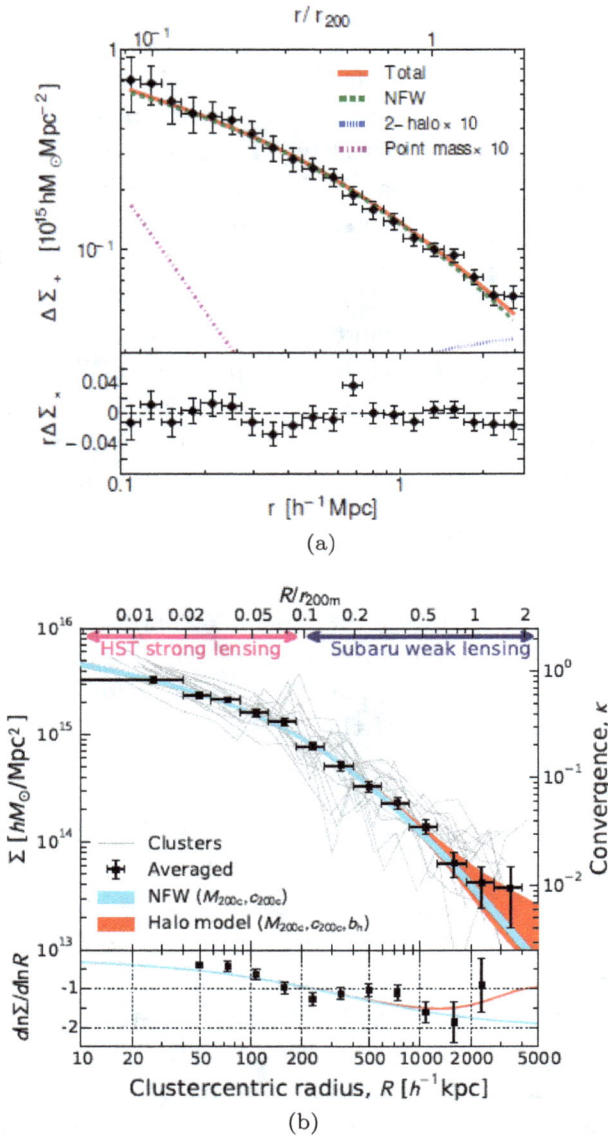

Fig. 3.15. Averaged radial mass profile of LoCUSS sample (a) and CRASH sample (b).

shows the stacked shear profile over the radial range $100\,h^{-1}$ kpc $<$ $r < 2.8\,h^{-1}$ Mpac. This shows that the profile is well described by the predicted NFW profile with parameters $M_{\rm vir} = 7.19^{+0.53}_{-0.50} \times 10^{14}h^{-1}M_\odot$ and $c_{\rm vir} = 5.41^{+0.49}_{-0.45}$ with less than 10% statistical error. The concentration parameter obtained from stacked lensing profile is broadly in line with theoretical predictions (e.g. Bhattacharya *et al.* [165]). As Fig. 3.13 (a) shows, there is no evidence of departures from the NFW profile.

Okabe and Smith have improved the selection of background galaxies and confirmed the above result [158]. They showed that their new LoCuSS mass measurements are consistent with CCCP and CLASH, but is in tension with WtG at $\sim \sigma - 2\sigma$ significance.

## CRASH survey

CRASH (Cluster Lensing and Supernova Survey wih Hubble) [208] is to study dark matter in distant galaxies, cluster galaxy evolution and the geometry of the universe by observing 25 X-ray selected massive galaxy cluster (Zitrin *et al.* [209]; Coe *et al.* [210]; Umetsu *et al.* 2012) by Hubble space telescope.

CLASH survey carried out extremely deep imaging in 16 wavelength bands of the central regions of 25 massive galaxy clusters from 2010 to 2013 using the HST. The combination of HST strong lensing results and Subaru weak lensing results for CRASH clusters eas carried out by Umetsu *et al.* [145] using the Bayesian approach developed by Umetsu [211]. In the Bayesian framework, the lensing signal is described by a vector of parameters $\kappa(R_i)$ containing the binned convergence profile. In this framework, multiple lensing constraints, including tangential reduced shear, magnification bias, and strong-lensing enclosed mass measurements, are combined a posterior, in the form of azimuthally averaged profiles, to reconstruct the binned convergence profile $\kappa(R_i)$. Figure 3.13(b) is the resulting the average surface density $\langle \Sigma(r) \rangle$ for the CLASH sample. The figure shows that the mass distribution of the CLASH sample is consistent with the universal NFW model over a two-order radial range of $40\,h^{-1}\mathrm{kpc} < r < 4000\,h^{-1}\mathrm{kpc}$.

In the central region of a galaxy cluster ($r \leq 20\,h^{-1}\,\mathrm{kpc}$), the influence of baryons in giant elliptical galaxies cannot be ignored, but in the lens observation region ($r \leq 40\,h^{-1}\,\mathrm{kpc}$), dark matter dominates the matter distribution. Applying the NFW model to $\langle\Sigma\rangle(r)$ distribution of this CLASH sample, Umetsu *et al.* [145]. estimated the halo mass $M_{200} \simeq 1.4 \times 10^{14} M_{\odot}$ and concentration parameter $c_{200} = 3.95 \pm 0.35$. The concentration parameter agrees within error with the theoretical predictions of the standard CDM model at the mean redshift of the CLASH sample, $\langle z \rangle = 034$. For another approach to combine the strong lensing and weak lensing, see Merten *et al.* [216].

An accurate estimate of cluster mass is crucially important for utilizing clusters as cosmological probes [144, 146, 147]. Conventionally, the total cluster mass is determined from projected measurements assuming spherical symmetry. However, N-body simulations in the LCDM model indicate that cluster halos are predicted to be non-spherical, with a preference for prolate shapes [148–151]. Oguri *et al.* compared the observed shapes of galaxy clusters using a sub sample of LoCuSS sample and showed that the mean ellipticity of the dark matter distribution is consistent with the prediction of the triaxial halo mode [152]. Chiu *et al.* has performed the three-dimensional modeling of intrinsic mass distribution using a subsample of CLASH sample [153]. They have presented a statistical three-dimension analysis of a sizable sample of high-mass galaxy clusters using high-quality weak and strong lensing data sets. Although clear evidence of a departure from spherical symmetry in their sample of 20 clusters was discovered, it was shown that the assumption of spherical symmetry is still well validated in determining the overall mass profile(such as concentration and mass) if the sample is free from orientation bias. The number of samples are still small to have a definite to say anything definite on the statistical properties of cluster and it is highly desirable to extend this type of analysis to on-going and near future survey.

## Dark matter subhalos in the coma cluster

According to the CDM structure formation scenario, less massive halos are accreted into more massive halos, which are then subsequently eroded by effects combined with tidal stripping and dynamical friction of the host halo, eventually becoming a smooth component. In this process, the central regions of subhalos have survived under the overdensity field, and constitute their population. Numerical simulations of, and analytic approaches to CDM based structure formation scenaric predict that subhalo mass functions at the intermediate and low mass scales follow a power law, $dn/dlnM_{sub} \propto M_{sub}^{-\alpha}$ with $\alpha \sim 0.9-1.0$ (e.g. Taylor and Babul [166–168]; Oguri and Lee [169]; van den Bosch et al. [170]; Diemand et al. [171]; De Lucia et al. [172]; Gao et al. [173]; Shaw et al. [174]; Angulo et al. [175]; Giocoli et al. [176]; Klypin et al. [177]; Gao et al. [178]; Wu et al. [179]).

Thus, the statistical properties of cluster subhalos such as mass function and spatial distribution provide us with useful information about the mass assembly history. Their measurement is the most stringent test of CDM predictions on scales of less than several Mpc. A characteristic feature of the subhalo mass function is also critically important to constrain the nature of dark matter, because it depends on the particle mass of dark matter. Furthermore, measurement of the correlation between member galaxies and subhalo masses sheds important insight on the physics of galaxy evolution associated with dark matter. Thus, it is important to measure the mass function directly from observations without assuming a relationship between dark matter and luminous matter and the dynamical state of the system.

Weak lensing observations of very nearby clusters ($z \leq 0.05$) are ideal for this purpose because their large apparent size enables us to easily resolve less massive subhalos inside the clusters and provides a correspondingly large number of background galaxies, which leads to low statistical errors and compensates for the low lensing efficiency to achieve a high S/N.

We show the results of a 4.1 deg$^2$ weak gravitational lensing survey of subhalos in the very nearby Coma cluster at redshift $z = 0.0236$ using the Subaru/Suprime-Cam (Okabe *et al.*, 2012). The observed area is about 80% within $r_{200}$ inside of which the mean density is 200 times the critical density at the cluster redshift. Note that one arcmin corresponds to $20 \, h^{-1}$ kpc for the cosmology with $\Omega_{m,0} = 0.27, \Omega_{\Lambda,0} = 0.73$ and $H_0 = 100 \, \mathrm{hkms^{-1}Mpc^{-1}}$.

Figure 3.16 shows the projected mass distribution with a smoothing scale of FWHM $= 4'$. and units of significance of $\nu = \kappa/\sigma_\kappa$. As shown in this figure, 32 subhalos were detected and their masses are measured in a model-independent manner, down to the order of $10^{-3}$ of the virial mass of the cluster. All of them are associated with a small group of cluster members. The mean distortion profiles stacked over subhalos show a sharply truncated feature well-fitted by a Navarro–Frenk–White (NFW) mass model with the truncation radius, as expected due to tidal destruction by the main cluster.

It is found that subhalo masses, truncation radii, and mass-to-light ratios decrease toward the cluster center. Figure 3.17 shows the subhalo mass function $dn/dlnM_{\mathrm{sub}}$ constructed by weak lensing mass measurements. It covers a range of two orders of magnitude in mass and is well described by a single power law

$$\frac{dn}{d \ln M_{\mathrm{sub}}} \propto M_{\mathrm{sub}}^{-\alpha} \qquad (3.4.48)$$

with $\alpha = 1.09^{+0.42}_{-0.32}$.

This agrees with the predicted value of $\sim 0.9$–$1.0$ from an N-body simulation based on the CDM scenario.

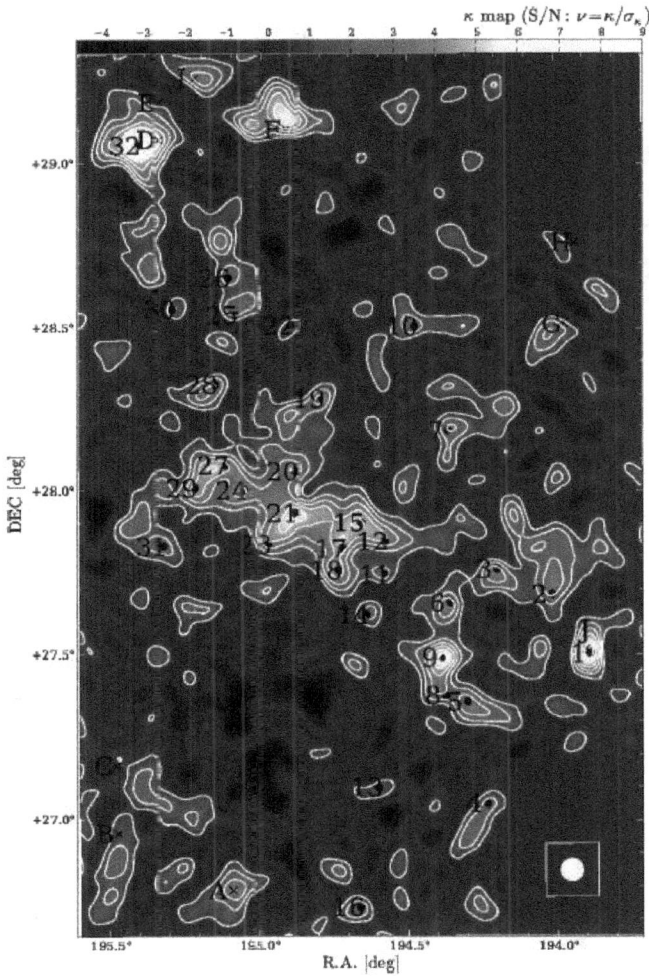

Fig. 3.16. Projected mass distribution of th Coma cluster.

## 3.4.5. *Cosmic shear*

The most challenging observation in weak lensing is cosmic shear, which is weak lensing due to the large scale structure of the universe. It attracted much attention because its signal depends on the expansion rate of the universe and the growth rate of structure. It

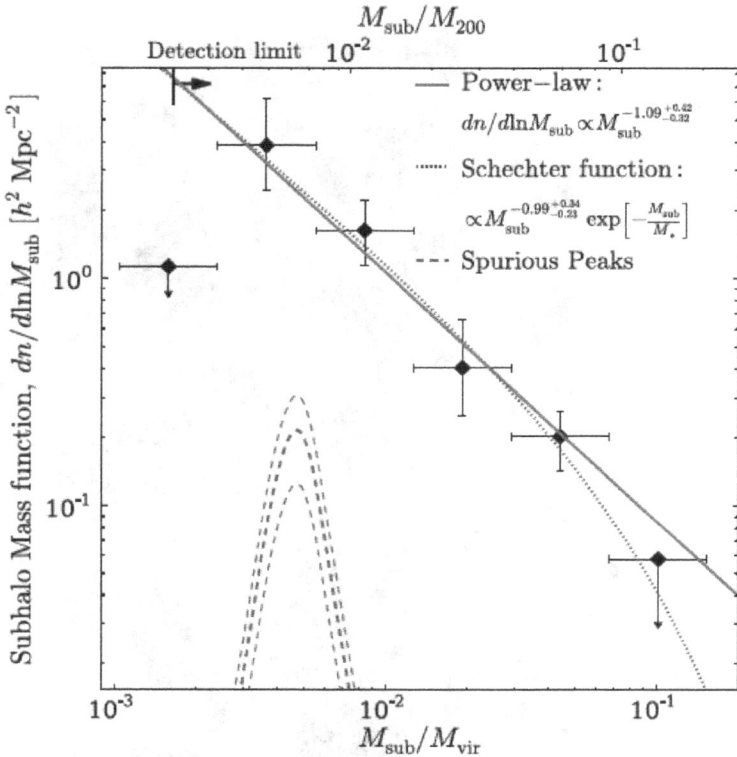

Fig. 3.17. Subhalo mass function of Coma Cluster.

therefore carries useful information about dark energy and the theory of gravity.

Cosmic shear has already been detected by several groups [43–46]. Subsequent detections have put useful constraints on the matter density parameter $\Omega_{m,0}$ and on the matter density spectrum normalization $\sigma_8$ (Bacon *et al.* [181]; Hoekstra, Yee and Gladders [182]; Bacon *et al.* [183]; Hamana *et al.* [121]). More recent results are found, for example, in Fu *et al.* [184] for the CFHTLS third data release; Schrabback *et al.* [185] for the Hubble Space Telescope COSMOS survey; Benjamin *et al.* [186] and Kilbinger *et al.* [188] for CFHTLens; and Jee *et al.* [189] for the Deep Lens Survey.

In particular, the Canada–France–Hawaii Telescope Lensing Survey (CFHTLenS) covers $154 \, \mathrm{deg}^2$ in five optical bands where the shapes of 4.2 million galaxies with $0.2 < z < 1.3$ (photometric redshift) are measured. Combining with WMAP7, BOSS and an HST distance-ladder prior on the Hubble parameter, it is found that $\Omega_{m,0} = 0.283 \pm 0.010$ and $\sigma_8 = 0.813 \pm 0.04$ for a flat $\Lambda$CDM model.

On-going and planned galaxy surveys aiming to observe cosmic shear including the Panoramic Survey Telescope & Rapid Response System (Pan-STARRS) [70], the Subaru Hyper Suprime Cam (HSC) Survey [190], the Dark Energy Survey (DES) [191], the Large Synoptic Survey Telescope (LSST) [192], the KIlo-Degree Survey (KIDS) [193] and Euclid [194]. However the signal is extremely small and contains many systematic errors such as inaccurate shape measurement, and inaccurate determination of redshifts. Also, we do not yet have a complete theoretical understanding of small scale clustering. Thus, it is essential to reduce these systematic errors down to a required level in order to place a useful constraint on the cosmological information (Huterer *et al.* [195]; Cropper *et al.* [196]; Massey *et al.* [197]). This is an ongoing and actively developing research field. Here we give an introduction to understand the physics of cosmic shear.

Cosmic shear observations measure the coherent distortion of background galaxies. This is related to the convergence of the matter density field and thus the matter power spectrum. In fact, there is a relationship between the shear and convergence:

$$\langle \hat{\gamma}(\mathbf{k})\hat{\gamma}(\mathbf{k}) \rangle = \langle \hat{\kappa}(\mathbf{k})\hat{\kappa}(\mathbf{k}') \rangle = (2\pi)^2 C_\kappa(k)\delta_D(\mathbf{k} - \mathbf{k}') \quad (3.4.49)$$

where the quantities with a "hat" are Fourier transforms of the corresponding quantities, and $C(k)$ is the lensing power spectrum of the shear (convergence). As easily imagined, the power spectrum is related to the matter power spectrum, and thus the observation of the cosmic shear provides a method to measure the evolution of the matter power spectrum.

We derive the basic equation for cosmic shear. Since the cosmic shear is the weak lensing phenomena by the continuous distribution of the three-dimensional gravitational potential $\Psi$, the lensing potential is obtained by integrating the potentil along the line of sight up to the source position.

$$\psi(\boldsymbol{\theta}, \chi_s) = \frac{2}{c^2} \int_0^{\chi_s} d\chi \frac{r(\chi_s - \chi)}{r(\chi_s)r(\chi)} \Psi(r(\chi)\boldsymbol{\theta}, \chi) \qquad (3.4.50)$$

where $\boldsymbol{\theta}$ is the angular vector on sky. In observation, we will observe huge number of background galaxies to obtain the lensing information so that we also integrated over source (vakground galaxies) distribution $p_s(\chi_s)$ as

$$\psi(\boldsymbol{\theta}) = \frac{2}{c^2} \int_0^{\chi_h} d\chi_s p_s(\chi_s) \int_0^{\chi_s} d\chi \frac{r(\chi_s - \chi)}{r(\chi_s)r(\chi)} \Psi(r(\chi)\boldsymbol{\theta}, \chi)$$

$$= \frac{2}{c^2} \int_0^{\chi_h} d\chi \frac{q_s(\chi)}{r(\chi)} \Psi(r(\chi), \chi) \qquad (3.4.51)$$

where $\chi_h$ is the comoving distance of the present horizon and we defined $q_s$ as follows.

$$q_s(\chi) = \int_\chi^{\chi_h} d\chi' p_s(\chi') \frac{r(\chi' - \chi)}{r(\chi')} \qquad (3.4.52)$$

the probability distribution of background galaxies is normalized as

$$\int_0^{\chi_h} d\chi \, p_s(\chi) = 1 \qquad (3.4.53)$$

Then the convergence is written as

$$\kappa(\boldsymbol{\theta}) = \frac{1}{2} \Delta_\theta \psi = \frac{3H_0^2 \Omega_{m,0}}{2c^2} \int_0^{\chi_h} d\chi \frac{q_s(\chi)r(\chi)}{a(\chi)} \delta_m(r(\chi)\boldsymbol{\theta}, \chi) \quad (3.4.54)$$

where we have used the cosmological version of Poisson equation.

$$\Delta\Psi = 4\pi Ga^2 \rho_m \delta_m = \frac{3}{2} H_0^2 \Omega_{m,0} \frac{\delta_m}{a} \qquad (3.4.55)$$

As seen from the above equation, the dependence of the cosmological parameters appears in two parts, in the comoving distance and in the evolution of the density contrast. The latter depends not only on the global geometry of the universe but also on the local gravitational force, and thus cosmic shear is in principle able to distinguish between dark energy and an alternative theory of gravity as the origin of the accelerated expansion.

The power spectrum for shear (convergence) may be written in terms of the matter power spectrum $P_\delta$ as follows:

$$C_\kappa(\ell) = \frac{9H_0^4\Omega_m^2}{4} \int_0^{\chi_h} d\chi \frac{q_s^2(\chi)}{a^2(\chi)} P_m\left(k = \frac{\ell}{r(\chi)}; \chi\right) \quad (3.4.56)$$

This is the desired relationship between the power spectrum of shear and the matter power spectrum. In the current cosmic shear measurement, a sample of background galaxies is devided in several subsamples ($a = 1, 2, ..., N$) with some redshift range. The probability distribution of each sample $p_s^{(a)}$ is specified by measuring the redshift of galaies. The information of the evolution of matter density constrat is obtained by mezsureing correlations between different subsambles. This method of measurement is called tomographic method. The power spectrum of the convergence between sabsample (a) and (b) is obtained by the correlation function between sabsamble (a) and (b) and written as follows

$$C_\kappa^{(a)(b)}(\ell) = \frac{9}{4}\Omega_{m,0}^2 \left(\frac{H_0}{c}\right)^4 \int_0^{\chi_h} d\chi \frac{q_s^{(a)}(\chi)q_s^{(b)}(\chi)}{a^2(\chi)} P_m\left(k = \frac{\ell}{r(\chi)}, \chi\right)$$

$$(3.4.57)$$

where $q_s^{(a)}$ is kernel function definned from sabsample (a) using eq. (3.4.52).

In cosmic shear observations, the number of background galaxies is huge and the spectrocsopic determination of redshift is not practical. Therefore, the redshifts are determined by photometric information using several bands. The determination of photometric redshift brings about the most serious ambiguity in the current cosmic shear measurement.

## 2-point correlation function of shear

In the actual cosmic shear measurement, shear power spectrum or 2-point correlation function of shear is measured from galaxy correlation function. in the following way. Suppose a $n$th galaxy at the position $\boldsymbol{\theta}_n$ on sky has an intrinsic (complex )ellipticity $\epsilon^{(s)}(\boldsymbol{\theta}_n)$. Then the observed (complex) ellipticity $\epsilon^{(obs)}$ is related with complex shear $\gamma$ as follows.

$$\epsilon^{(obs)}(\boldsymbol{\theta}_n) = \epsilon^{(s)}(\boldsymbol{\theta}_n) + 2\gamma(\boldsymbol{\theta}_n) \tag{3.4.58}$$

Then we measure the correlation between two galaxies with a fixed angular distance $\theta = |\boldsymbol{\theta}_n - \boldsymbol{\theta}_m|$ and average over many such correlations.

$$\langle \epsilon^{(obs)}(\boldsymbol{\theta}_n)\epsilon^{(obs)}(\boldsymbol{\theta})\rangle = \sigma_\epsilon^2 \delta_{nm} + 4\langle\gamma\gamma^*\rangle(\theta) \tag{3.4.59}$$

Here, we assumed that the intrinsic orientation of galaxies are random and $\sigma_\epsilon^2 = \langle|\epsilon^{(s)}|^2\rangle$ is its dispersion. In this way, the measurement of galaxy-galaxy correlation function gives the measurement of the 2-point correlation function of shear.

In actual cosmic shear measurement, the following two point correlation functions are used:

$$\xi_\pm(\theta) = \langle\gamma_t\gamma_t\rangle(\theta) \pm \langle\gamma_\times\gamma_\times\rangle(\theta) \tag{3.4.60}$$

$$\xi_\times(\theta) = \langle\gamma_t\gamma_\times\rangle(\theta). \tag{3.4.61}$$

Since $\gamma_t \to \gamma_t$ and $\gamma_\times \to -\gamma_\times$ under the parity transformation, $\xi_\times$ should vanish identicaly. Thus it can be used as an indicator of measureent error.

These correlation functions may be expressed in terms of the convergence power spectrum. For example,

$$\xi_+ = \int \frac{d^2\ell}{(2\pi)^2} \int \frac{d^2\ell'}{(2\pi)^2} e^{i\boldsymbol{\ell}\cdot(\boldsymbol{\theta}'+\boldsymbol{\theta})-i\boldsymbol{\ell}'\cdot\boldsymbol{\theta}'} \langle \gamma(\boldsymbol{\ell})\gamma^*(\boldsymbol{\ell}')\rangle \qquad (3.4.62)$$

$$= \int \frac{\ell d\ell}{2\pi} J_0(\ell\theta) C_\kappa(\ell) \qquad (3.4.63)$$

where $J_n(x)$ is the nth order Bessel function of first kind.

$$J_n(x) = \frac{1}{2\pi} \oint d\phi\, e^{in\phi - ix\sin\phi} \qquad (3.4.64)$$

Similarly, one can derive the expression for $\xi_-$.

$$\xi_- = \int \frac{d\ell\ell}{2\pi} J_4(\ell\theta) C_\kappa(\ell) \qquad (3.4.65)$$

Using the above expression, we can express the angular power spectrum in terms of the 2-point correlation function as follows.

$$C_\kappa(\ell) = 2\pi \int d\theta\theta\xi_+(\theta) J_0(\ell\theta) = 2\pi \int d\theta\theta\xi_-(\theta) J_4(\ell\theta) \quad (3.4.66)$$

where we use the orthogonality of the Bessel functions

$$\int_0^\infty dx x J_n(ux) J_n(vx) = \frac{1}{u}\delta_D(u - v) \qquad (3.4.67)$$

In practical measurements of the two-point correlation function, we need to construct an unbiased estimator and a covariance matrix in order to compare observables with theory. We ask the reader to consult relevant references such as Schneider *et al.* [201] and Semboloni *et al.* [202] for detail.

Figure 3.18 shows the results of the cosmic shear observation using the 3.6 m Canada–France–Hawaii Telescope located on the summit of Mauna Kea on the Big Island of Hawaii. In this observation, a sky area spanning 154 square degrees was imaged in five wavelength bands, and the number of background galaxies used for analysis reached approximately 4.2 million. The figure shows the two-point

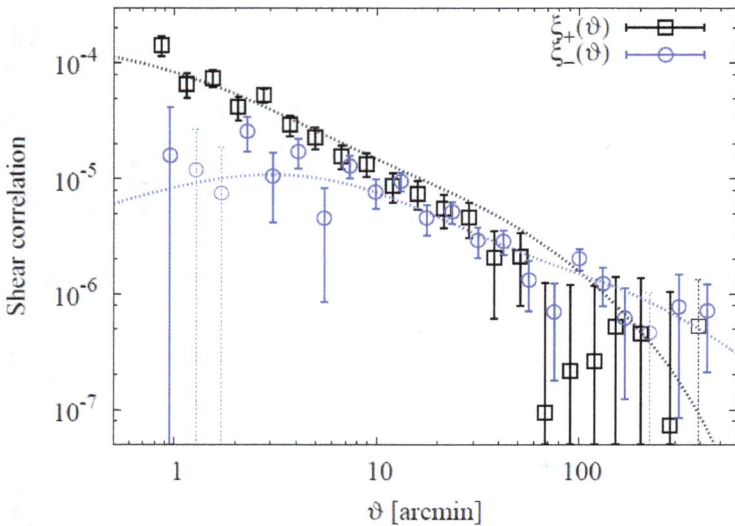

Fig. 3.18. The 2-point correlation functions of cosmic shear observed by Canada-France-Hawai Telescope.

correlation functions $\xi_+(\theta)$ and $\xi_-(\theta)$ of the cosmic shear measured over an angular scale of 0.8 arcmin to 350 arcmin. As explained above, the cosmic shear signal depends on the radial distance of the background galaxy, and thus accurate redshift information of the background sample is essential. However, spectroscopic observation takes a huge amount of time and not used in this observation. Instead the redshift of the background galaxy is estimated by multi-wavelength photometric data.

The ongoing large-scale multi-band photometric surveys which have weak lensing among their primary science targets include the Kilo-Degree Survey [218], the Dark Energy Survey [219], and the Hyper Suprime-Cam survey [220]. In these surveys the background galaxies are devided by several redshift bins and the tomographic measuremnts have been performed.

Figure 3.19 shows the two-point correlation functions(2PCFs) constructed by the three-year data of HSP–SSP. HSP is the

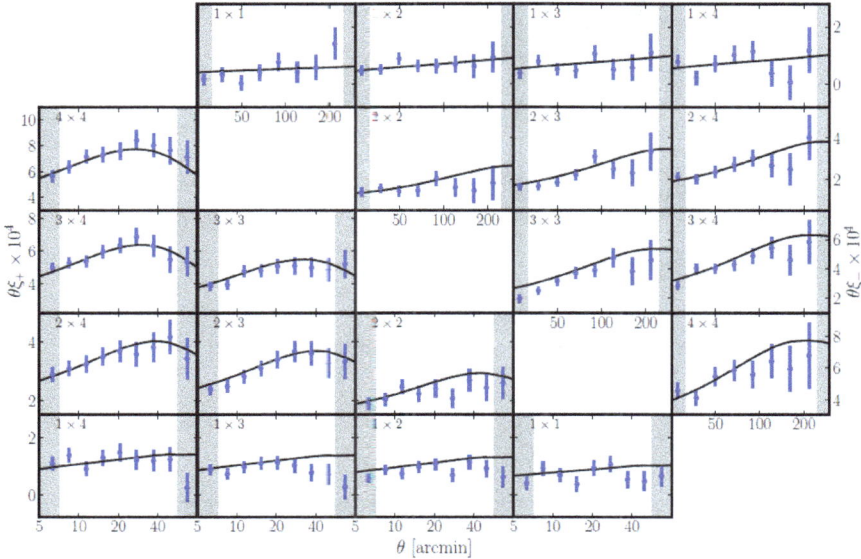

Fig. 3.19. Tne 2PCFs (four autocorrelations and six correlations between 4 tomographic redshift bins labeled with 1-4.) obtained by HSP–SSP 3 year results. These functions are constructed by the data over $5.3 < \theta < 56.6$[arcmin] for $\xi_+$ and $23.2 < \theta < 248$ [arcmin] for $\xi_-$ The solid lines are best fit model of th fiducial analysis. See detail in X. Li *et al.* [218].

ultra-wide-field prime focus came of Subaru telescope. one of the purpose of SSP (Subaru Strategic Program) is to measure accurately the cosmic shear to study the nature of dark energy. HSC–SSP carried out multi band imaging and shape measurements of a huge number of galaxies with four redshift bins ranging from 0.3 to 1.5 over approximately 1100 squre degree.

The three-year data covered an area of 416 square degrees, and the weak gravitational lensing analysis used observation data of about 25 million background galaxies with i-band magnitudes up to about 26 AB magnitude with an effective galaxy number density of 15 arcmin$^{-2}$ over the four redshift bins. The correlation functions are determined from the correlation of these subsamples. Figure 3.20 shows the constraints on the matter density parameter $\Omega_{m,0}$ and

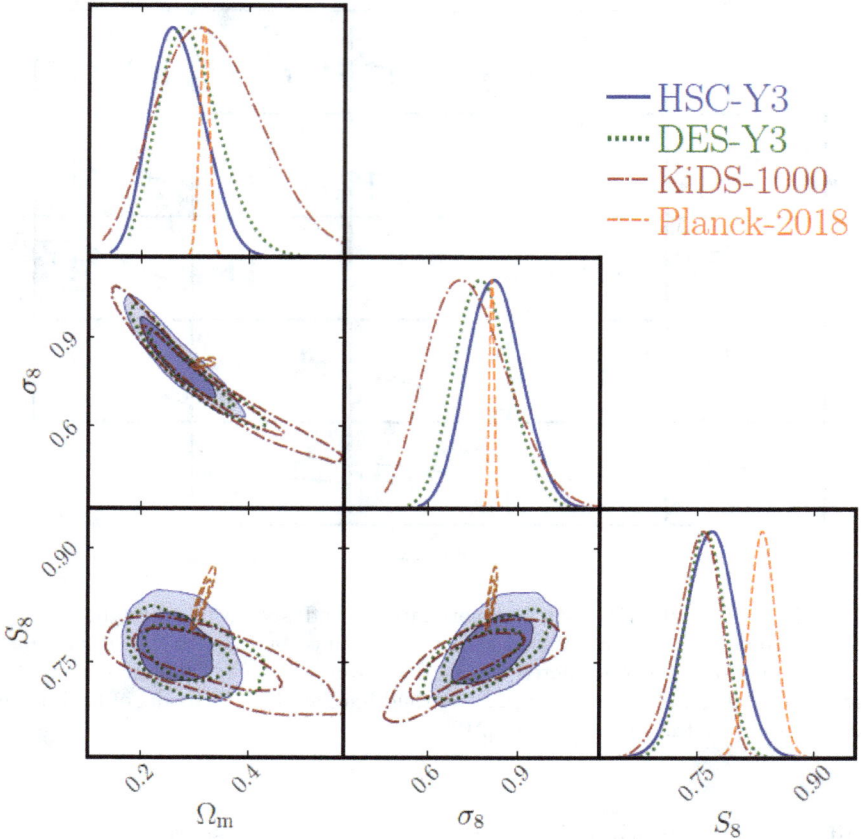

Fig. 3.20.　Costraints on $\Omega_m$, $\sigma_8$ and $C_8$ from various experiments.

structure formation parameter $\sigma_8$r using the obtained cosmic shear results obtained by the three-year data of HSC [221], The figure shows $\Omega_{m,0} = 0.256^{+0.056}_{-0.044}$ and $S_8 = \sigma_8\sqrt{\Omega_{m,0}/0.3} = 0.769^{+0.031}_{-0.034}$. These values are consistent with results reported by the ongoing surveys:KiDS-1000, DES -Y3 [219]. However, the weak-lensing constraints on $S_8$ is $\sim 2\sigma$ lower than the constraint from Planck-2018 result [214]. This disagreement is called $S_8$ tension. Concerning the origin of the accelerated expansion of the universe, any definite constraints are not yet obtained. Currently, an international large-scale galaxy survey is underway to measure the cosmic shear more precisely, and surveys using space telescopes such as the Euclid and Nancy Grace Roman telescopes are underway. It is expected that we are going to know more details on the nature of dark energy in near future.

# References

1. D. Valls-Gabaud, *Albert Einstein Century Int. Conf.* **861** (2006) 1163.
2. J. Soldner, *Astronomishes Jahrbuch fur* **1804** (1801) 161.
3. G. B. Caminha, S. H. Suyu, A. Mercurio, *et al. Astro. & Astrophys.* **666** (2022) L9.
4. O. Chwolson, *Astronomische Nachrichten* **221** (1924) 329.
5. A. Einstein, *Science* **84** (1936) 506.
6. F. Link, *Comptes Rendus Acad. Sci.* **202** (1936) 917.
7. F. Link, *Bull. Astron.* **10** (1937) 73.
8. F. Zwicky, PR **51** (1937) 290.
9. S. Refsdal, *Mon. Not. R. Astron. Sco.* **128** (1964) 307.
10. S. Liebes, *Phys. Rev.* **133** (1964) 835.
11. D. Walsh, R. F. Carswell and R. J. Weymann, *Nature* **279** (1979) 381.
12. A. Lewis and A. Challinor, *Phys. Rept.* **429** 1 (2006).
13. T. Kundić *et al.*, *Astrophys. J.* **482** (1997) 75.
14. R. Lynds and V. Petrosian, *Bull. Am. Astron. Soc.* **18** (1986) 1014.
15. A. Hoag, *Bull. AAS*, **13** (1981) 799.
16. G. Soucail, B. Fort, Y. Mellier and J. P. Picat, *AA* **172** (1987) L14.
17. J. Hewitt, *Nature* **333** (1988), 537–540.
18. J. N. Hewitt, E. L. Turner, D. P. Schneider, B. F. Burke and G. I. Langston, *Nature* **333** (1988) 537.
19. J.-P. Kneib, R. S. Ellis, I. Smail, W. J. Couch and R. M. Sharples, *Astrophys. J.* **471** (1996) 643.
20. T. Futamase and S. Yoshida, *Prog. Theor. Phys.* **105** (2001) 887–891.
21. K. Yamamoto and T. Futamase, *Progr. Theor. Phys.* **105** (2001) 707–716.
22. G. Golse, J.-P. Kneib and G. Soucail, *Astron&Astrophts*, **387** (2002) 788–803.
23. M. Managhetti *et al. Mor.. Not. Roy. Astrono. Soc.* **362** (2005) 1301–1310.

24. E. Jullo *et al.*, *Science* **329** (2010) 924.
25. R. S. Ellis, arXiv:0109294.
26. J.-P. Kneib, R. S. Ellis, M. R. Santos and J. Richard, *Astrophys. J.* **607** (2004) 697.
27. J. Richard *et al.*, *Astrophys. J.* **685** (2008) 705.
28. R. J. Bouwens *et al.*, *Astrophys. J.* **690** (2009) 1764.
29. D. Coe *et al.*, *Astrophys. J.* **762** (2013) 32.
30. M. Bradač *et al.*, *Astrophys. J.* **652** (2006) 937.
31. M. Bradač *et al.*, *Astrophys. J.* **706** (2009) 1201.
32. E. Jullo *et al.*, *New J. Phys.* **9** (2007) 447.
33. E. Jullo and J.-P. Kneib, *Mon. Not. R. Astron. Soc.* **395** (2009) 1319.
34. J. Richard *et al.*, arXiv:1405.3303.
35. E. L. Turner, J. P. Ostriker and J. R. Gott, *Astrophys. J.* **284** (1984) 1.
36. E. Turner, *Astrophys. J.* **365** (1990) L43.
37. M. Fukugita, T. Futamase and M. Kasai, *Mon. Not. R. Astron. Soc.* **246** (1990) 24P.
38. J. A. Tyson, R. A. Wenk and F. Valdes, *ApJL* **349** (1990) L1.
39. M. Kasai and T. Futamase, *Progr. Theor. Exper. Phys* **2019**, 7, 073E01.
40. N. Kaiser and G. Squires, *Astrophys. J.* **404** (1993) 441.
41. N. Kaiser, G. Squires and T. Broadhurst, *Astrophys. J.* **449** (1995) 460.
42. N. Okabe, G. P. Smith, K. Umetsu, M. Takada and T. Futamase, *ApJL* **769** (2013) L35.
43. L. van Waerbeke *et al.*, *A&A* **358** (2000) 30.
44. D. J. Bacon, A. R. Refregier and R. S. Ellis, *Mon. Not. R. Astron. Soc.* **318** (2000) 625.
45. N. Kaiser, G. Wilson and G. A. Luppino, arXiv:astro-ph/0003338.
46. R. Maoli *et al.*, *A&A* **368** (2001) 766.
47. L. Van Waerbeke *et al.*, *A&A* **393** (2002) 369.
48. D. Wittman *et al.*, *Nature* **405** (2000) 143.
49. B. Paczynski *et al.*, *Astrophys. J.* **435** (1994) L113.
50. P. Schneider, J. Ehlers and E. E. Falco, *Gravitational Lenses*, Vol. XIV, Springer-Verlag, Berlin, 1992.
51. R. D. Blandford and R. Narayan, **30** (1992) 311.
52. S. Refsdal and J. Surdej, *Rep. Prog. Phys.* **57** (1994) 117.
53. R. Narayan and M. Bartelmann, arXiv:astro-ph/9606001.
54. T. Futamase, *Mon. Not. R. Astron. Soc.* **237** (1989) 187.
55. L. L. R. Williams and G. F. Lewis, *Mon. Not. R. Astron. Soc.* **294** (1998) 299.
56. T. Futamase, M. Hattori and T. Hamana, *ApJL* **508** (1998) L47.
57. A. Zitrin and T. Broadhurst, *ApJL* **703** (2009) L132.

58. M. Bartelmann, *AA* **303** (1995) 643.

59. C. O. Wright and T. G. Brainerd, arXiv:9908213.

60. N. A. Grogin and R. Narayan. *Astrophys. J.* **464** (1996) 92.

61. R. D. Blandford and C. S. Kochanek, *Dark Matter in the Universe*, in J. Bahcall, T. Piran and S. Weinberg (eds.), World Scientific, Springer, p. 133.

62. M. Hattori, J. Kneib and N. Makino, *Progress Theor. Phys. Suppl.* **133** (1999) 1.

63. C. S. Kochanek, in *Proc. 33rd Saas-Fee Advanced Course*, G. Meylan, P. Jetzer and North P. (eds.), Springer-Verlag, Berlin.

64. J. P. Kneib and P. Natarayan. *Astron. Astrophys. Rev.* **19** (2011) 47.

65. S. Mao and P. Schneider, *Mon. Not. R. Astron. Soc.* **295** (1998) 587.

66. M. Chiba, *Astrophys. J.* **565** (2002) 17.

67. A.G. Riess and L. Brueval, Proceedings IAU Symp. No.376, Edited by R.de Grijs, P.A. Whitelock & M. Catelan, (2024), Cambridge University Press.

68. S. H. Suyu *et al.*, *Astrophys. J.* **766** (2013) 70.

69. S. H. Suyu *et al.*, *ApJL* **788** (2014) L35.

70. N. Kaiser, W. Burgett and K. Chambers *et al.*, *Proc. SPIE* **7733** (2010) 77330.

71. M. Oguri and P. J. Marshell, *MNRAS* **405** (2010) 2579.

72. E. V. Linder, *Phys. Rev. D* **84** (2011) 123529.

73. H. Dahle, S. J. Maddox and P. B. Lilje, *Astrophys. J.* **435** (1994) L79.

74. P. Fisher, G. Bernstein, G. Rhee and J. A. Tyson, *AJ* **113** (1997) 521.

75. A. T. Lefor, T. Futamase and M. Akhlaghi, *NewAR* **57** (2013) 1.

76. P. Saha and L. L. R. Williams, *AJ* **127** (2004) 2604.

77. R. B. Wayth and R. L. Webster, *Mon. Not. R. Astron. Soc.* **372** (2006) 1187.

78. D. Coe *et al.*, *Astrophys. J.* **681** (2008) 814.

79. J. Liesenborgs *et al.*, *Mon. Not. R. Astron. Soc.* **386** (2008) 307.

80. M. Oguri, *PASJ* **62** (2010) 1017.

81. C. R. Keeton, arXiv:astro-ph/0102340.

82. J. Rigby, M. J. Perrin. M. McElwain, *et al.* (2022), arXiv e-prints, arXiv:2207.05632.

83. E. L. Turner, *ApJL* **365** (1990) L43.

84. C. S. Kochanek, *Astrophys. J.* **466** (1996) 638.

85. E. O. Ofek, H.-W. Rix and D. Maoz, *Mon. Not. R. Astron. Soc.* **343** (2003) 639.

86. K.-H. Chae and S. Mao, *ApJL* **599** (2003) L61.

87. A. Matsumoto and T. Futamase, *Mon. Not. R. Astron. Soc.* **384** (2008) 843.

88. K.-H. Chae, *Mon. Not. R. Astron. Soc.* **402** (2010) 2031.

89. P. R. Capelo and P. Natarajan, *New J. Phys.* **9** (2007) 445.

90. M. Oguri *et al.*, *AJ* **143** (2012) 120.

91. M. Bernardi *et al.*, *Mon. Not. R. Astron. Soc.* **404** (2010) 2087.

92. S. M. Croom *et al.*, *Mon. Not. R. Astron. Soc.* **399** (2009) 1755.

93. N. Kaiser, *Astrophys. J* **537** (2000) 555.

94. J. Rhodes, A. Refregier and E. J. Groth, *Astrophys. J.* **536** (2000) 79.

95. C. M. Hirata and U. Seljak, *Mon. Not. Roy. Astr. Soc.* **353** (2003) 459.

96. Y. Okura and T. Futamase, *Astroophys. J.* **699** (2009) 143.

97. J. Zuntz, T. Kacprzak, L. Voigt *et al.*, *Mon. Not. Roy. Astr. Soc.* **434** (2013) 1604.

98. L. Miller, C. Heymans, T. D. Kitching *et al. Mon. Not. Roy. Astr. Soc.* **429** (2013) 2858.

99. T. D. Kitching, S. T. Balan, S. Bridle *et al. Mon. Not. Roy. astr. Soc.* **423** (2012) 3163.

100. R. Mandelbaum, *Ann. REv. Astron. Astrophys.* **56** (2018) 393.

101. S. Cao and Z.-H. Zhu, *A&A* **538** (2012) A43.

102. J. Tyson, R. A. Wenk and F. Valdes, *Astrophys. J.* **349** (1990) L1.

103. N. Kaiser and G. Squires, *Astrophys. J.* **404** (1993) 441.

104. P. Schneider and C. Seitz, *AA* **294** (1995) 411.

105. T. Brainerd, R. D. Blandford and I. Smail, *Astrophys. J.* **466** (1996) 623.

106. M. J. Huudson, S. D. Gwyn, H. Dahle and N. Kaiser, *Astrophys. J.* **503** (1998) 531.

107. D. Clowe, G. A. Luppino, N. Kaiser, J. P. Henry and I. M. Gioia, *Astrophys. J.* **498** (1998) L61.

108. H. Hoekstra, M. Franx, K. Kuijken and P. G. van Dokkum, *Mon. Not. R. Astron. Soc.* **333** (2002) 911.

109. H. Dahle *et al.*, *ApJS* **139** (2002) 313.

110. E. S. Cypriano, L. Sordre, J.-P. Kneib and L. E. Campusano, *Astrophys. J.* **613** (2004) 95.

111. M. Bartelmann and P. Schneider, *Phys. Rep.* **340** (2001) 291.

112. K. Umetsu, M. Tada and T. Futamase, *PTP Suppl.* **133** (1999) 53.

113. T. Broadhurst *et al.*, *Astrophys. J.* **619** (2005) L143.

114. E. Medezinski *et al.*, *Astrophys. J.* **663** (2007) 717.

115. K. Umetsu and T. Broadhurst, *Astrophys. J.* **684** (2008) 177.

116. K. Umetsu *Astrophys. J* **769** (2013) 13.

117. M. J. Hudson, S. D. J. Gwyn, H. Dahle and N. Kaiser, *Astrophys. J.* **503** (1998) 531.

118. H. Hoekstra, M. Franx, K. Kuijken and G. Squires, *Astrophys. J.* **504** (1998) 636.

119. T. Erben, L. van Waerbeke, E. Bertin, Y. Mellier and P. Schneider, *A&A* **366** (2001) 717.
120. M. Hetterscheidt *et al.*, *A&A* **468** (2007) 859.
121. T. Hamana *et al.*, *Astrophys. J.* **597** (2003) 98.
122. S. Seitz and P. Schneider, *AA* **305** (1998) 383.
123. L. van Waerbeke and Y. Mellier, arXiv:astro-ph/0305089.
124. M. Takada and M. White, *Astrophys. J.* **601** (2004) L1.
125. C. M. Hirata and U. Seljak, *PRD* **70** (2004) 063526.
126. G .P. Smith, *et al.*, *Mon. Not. Roy. Astr. Soc* **456** (2016) L74.
127. A. Mahdavi, H. Hoekstra. A. Babul *et al.*, *Astrophys. J* **767** (2013) 116.
128. H. Hoekstra, P. Herbonnet, A. Muzzin *et al.*, *Mon. Not. Roy. astr. Soc* **449** (2015) 685.
129. D. E. Applegate *et al.*, *Mon. Not. Roy. Astr. Soc* **439** (2014) 48.
130. A. vonder Linden *et al.*, *Mon. Not. Roy. Astr. Soc.* **430** (2014) 2.
131. S. Bridle *et al.*, *Ann. Appl. Stat.* **3** (2009) 6.
132. S. Bridle *et al.*, *Mon. Not R. Astron. Soc.* **405** (2010) 2044.
133. T. D. Kitching, A. F. Heavens and L. Miller, *Mon. Not. R. Astron. Soc.* **413** (2011) 2923.
134. T. D. Kitching *et al.*, *Mon. Not. R. Astron. Soc.* **423** (2012) 3163.
135. T. D. Kitching *et al.*, *ApJS* **205** (2013) 12.
136. C. Heymans *et al.*, *Mon. Not. R. Astron. Soc.* **368** (2006) 1323.
137. R. Massey *et al.*, *Mon. Not. R. Astron. Soc.* **376** (2007) 13.
138. T. D. Kitching *et al.*, *Mon. Not. R. Astron. Soc.* **423** (2012) 3163.
139. A. Leonard, D. M. Goldberg, J. L. Haaga and R. Massey, *Astrophys. J.* **666** (2007) 51.
140. A. Leonard, L. J. King and D. M. Goldberg, *Mon. Not. R. Astron. Soc.* **413** (2011) 789.
141. B. Cain, P. L. Schechter and M. W. Bautz, *Astrophys. J.* **736** (2011) 43.
142. M. Viola, P. Melchior and M. Bartelmann, arXiv:1107.3920.
143. B. Rowe *et al.*, arXiv:1211.0966.
144. S. Bocquet, A. Saro, J. Mohr *et al. Astrophys. J.* **799** (2015) 32.
145. K. Umetsu, A. Zitrin, D. Gruen, J. Merten, M. Donahue and M. Postman, *Astrophys. J.* **821** (2016) 116.
146. A. B. Mantz, A. van del Linden, S.W. Allen *et al.*, *Mon. Not. Roy. Astr. Soc* **446** (2015) 2205.
147. T. Se Haan, B. Benson, L. Bleen, *et al. Astrophys. J.* **832** (2016) 95.
148. C. S. Frenk, S. D. White, M. Davis and G. Efstathiou, *Astrophys. J.* **327** (1988) 507.
149. J. Dubinski and R. G. Carlberg, *Astrophys. J.* **378** (1991) 496.

150. M. S. Warren, P. J. Quinn, J. K. Salmon and W. H. Zurek, *Astrophys. J.* **399** (1992).
151. Y. P. Jing and Y. Suto., *Astrophys. J.* **574** (2002) 538.
152. M. Oguri, M. Takada, N. Okabe and G. P. Smith, *Mon. Not. Roy. Astr. Soc* **405** (2010) 2215.
153. I. N. Chiu, K. Umetsu, M. Sereno, *et al.*, *Astrophys. J.* **860** (2018) 126.
154. R. S. Levinson, *Publ. Astron. Soc. Jpn.* **125** (2013) 1474.
155. G. Fahlman, N. Kaiser, G. Squires and D. Woods, *Astrophys. J.* **437** (1994) 56.
156. D. E. Applegate *et al.*, *Mon. Not. R. Astron. Soc.* **439** (2014) 48.
157. H. Hoekstra *et al.*, *Mon. Not. R. Astron. Soc.* **427** (2012) 1298.
158. N. Okabe and G. P. Smith, *Mon. Not. R. Astron. Soc.* **461** (2016) 3794.
159. M. Oguri *et al.*, *Mon. Not. R. Astron. Soc.* **420** (2012) 3214.
160. N. Okabe and K. Umetsu, *Publ. Astron. Soc. Jpn.* **60** (2008) 345.
161. A. B. Newman *et al.*, *Astrophys. J.* **765** (2013) 24.
162. H. Ebeling *et al.*, *Mon. Not. R. Astron. Soc.* **301** (1998) 881.
163. H. Ebeling *et al.*, *VizieR Online Data Catalog* **730** (2000) 10881.
164. H. Böhringer *et al.*, *Astron. Astrophys.* **425** (2004) 367.
165. A. Bhattacharya *et al.*, *Astrophys. J.* **766** (2013) 32.
166. J. E. Taylor and A. Babul, *Mon. Not. R. Astron. Soc.* **348** (2004) 811.
167. J. E. Taylor and A. Babul, *Mon. Not. R. Astron. Soc.* **364** (2005) 515.
168. J. E. Taylor and A. Babul, *Mon. Not. R. Astron. Soc.* **364** (2005) 535.
169. M. Oguri and J. Lee, *Mon. Not. R. Astron. Soc.* **355** (2004) 120.
170. F. C. van den Bosch, X. Yang and H. J. Mo, *Mon. Not. R. Astron. Soc.* **340** (2003) 771.
171. J. Diemand, B. Moore and J. Stadel, *Mon. Not. R. Astron. Soc.* **352** (2004) 535.
172. G. De Lucia *et al.*, *Mon. Not. R. Astron. Soc.* **348** (2004) 333.
173. L. Gao, S. D. M. White, A. Jenkins, F. Stoehr and V. Springel, *Mon. Not. R. Astron. Soc.* **355** (2004) 819.
174. L. D. Shaw, J. Weller, J. P. Ostriker and P. Bode, *Astrophys. J.* **646** (2006) 815.
175. R. E. Angulo, C. G. Lacey, C. M. Baugh and C. S. Frenk, *Mon. Not. R. Astron. Soc.* **399** (2009) 983.
176. C. Giocoli, G. Tormen, R. K. Sheth and F. C. van den Bosch, *Mon. Not. R. Astron. Soc.* **404** (2010) 502.

177. A. A. Klypin, S. Trujillo-Gomez and J. Primack, *Astrophys. J.* **740** (2011) 102.
178. L. Gao *et al.*, *Mon. Not. R. Astron. Soc.* **425** (2012) 2169.
179. H.-Y. Wu, O. Hahn, R. H. Wechsler, P. S. Behroozi and Y.-Y. Mao, *Astrophys. J.* **767** (2013) 23.
180. K. Umetsu, E. Medezinski, M. Nonino *et al. Atrophys. J.* **795** (2014) 163.
181. D. J. Bacon, A. Refregier, D. Clowe and R. S. Ellis, *Mon. Not. R. Astron. Soc.* **325** (2001) 1065.
182. H. Hoekstra, H. K. C. Yee and M. D. Gladders, *Astrophys. J.* **577** (2002) 595.
183. D. J. Bacon, R. J. Massey, A. R. Refregier and R. S. Ellis, *Mon. Not. R. Astron. Soc.* **344** (2003) 673.
184. L. Fu *et al.*, *Astron. Astrophys.* **479** (2008) 9.
185. T. Schrabback *et al.*, in *Proceedings of the 6th International Heidelberg Conference*, H. V. Klapdor-Kleingrothaus of Geraint F. Lewis (ed.), World Scientific Publishing Co. 2008, pp. 260–273.
186. J. Benjamin *et al.*, *Mon. Not. R. Astron. Soc.* **431** (2013) 1547.
187. N. Okabe, T. Futamase, M. Kajisawa and R. Kuroshima, *Astrophys. J.* **784** (2014) 90.
188. M. Kilbinger *et al.*, *Mon Not. R. Astron. Soc.* **430** (2013) 2200.
189. M. J. Jee *et al.*, *Astrophys. J.* **765** (2013) 74.
190. The HSC collab., http://member.ip,u.jp/masahiro.takada/proposal_rv.pdf.
191. D. L. DePoy *et al.*, *SPIE* **7014** (2008) 70140E.
192. J. A. Tyson, D. M. Wittman, J. F. Hennawi and D. N. Spergel, in *Proc. 5th Int. UCLA Symp.* (2002).
193. G. Verdoes Kleijn *et al.*, arXiv:1112.0886.
194. Euclid Science Study. arXiv:0912.0914.
195. D. Huterer, M. Takada, G. Bernstein and B. Jain, *Mon. Not. R. Astron. Soc.* **366** (2006) 101.
196. M. Cropper *et al.*, *Mon. Not. R. Astron. Soc.* **431** (2013) 3103.
197. R. Massey *et al.*, *Mon. Not. R. Astron. Soc.* **429** (2013) 661.
198. J. A. Peacock and S. J. Dodds, *Mon. Not. R. Astron. Soc.* **280** (1996) L19.
199. A .G. Riesse *et al.*, *Astrophys. J. Lett.* **934** L7 (2022).
200. P. Schneider and M. Kilbinger, *Astron. Astrophys.* **462** (2007) 841.
201. P. Schneider, L. van Waerbeke, M. Kilbinger and Y. Mellier, *Acta Astronom.* **396** (2002) 1.
202. E. Semboloni *et al.*, *Mon. Not. R. Astron. Soc.* **375** (2007) L6.
203. M. Schmidt *et al.*, *Astrophys. J.* **507** (1998) 46.
204. S. Perlmutter *et al.*, *Nature* **391** (1998) 51.

205. J. F. Navarro, C. S. Frenk and S. D. White, *Astrophys. J.* **490** (1997) 493.
206. C. Seitz and P. Schneider, *Astron. Astrophys.* **318** (1997) 687.
207. H. Hoekstra, M. Franx and K. Kuijken, *Astrophys. J.* **532** (2000) 88.
208. M. Postman *et al.*, *ApJCS* **199** (2012) 25.
209. A. Zitrin *et al.*, *Astrophys. J.* **742** (2011) 117.
210. D. Coe *et al.*, *Astrophys. J.* **757** (2012) 22.
211. K. Umetsu, *Astrophys. J.* **795** (2014) 163.
212. K. Umetsu, T. Broadhurst, A. Zitrin, E. Medezinski and L. Hsu, *Astrophys. J.* **729** (2011) 127.
213. J.-P. Kneib and P. Natarajan, *Astron. Astrophys. Rev.* **19** (2011) 47.
214. N. Aghanim, Y. Akrami, M. Ashdown, J. Aumont, *et al. Astron. Astrophys* **641** (2020), A6.
215. K. Umetsu, *The Astron. Astrophysics. Rev* **28** (2020) 7.
216. J. Merten, M. Cacciato, M. Meneghetti, C. Mignone and M. Bartelmann, *Astron. Astrophys.* **500** (2009) 681.
217. K. C. Wong *et al.*, *Mon. Not. R. Astron. Soc.* **498** 1420 (2020).
218. M. Asgari, C.-A. Lin, B. Joachimi, B. Giblin, C. Heymans *et al.*, *Astron. Astrophys* **645**, A104 (2021).
219. T. Abbott, F. B. Abdalla, J. Aleksić, S. Allam *et al.*, *Mon. Not. R. Astron. Soc.* **460** 1270 (2016).
220. H. Aihara, N. Arimoto, R. Armstrong, S. Arnouts, N. A. Bahcall, S. Bickerton, J. Bosch, *et al.*, *Pub. Astron. Soc. Japan* **70**, S4 (2018).
221. X. Li, Zhang, S. Sugiyama, R. DAlal *et al.*, *Phys. Review D* **108** 123518 (2023).
222. R. DAlal, X. Li, A. Nicola, J. Zuntz *et al.*, *Phys. Review D* **108** 123519 (2023).